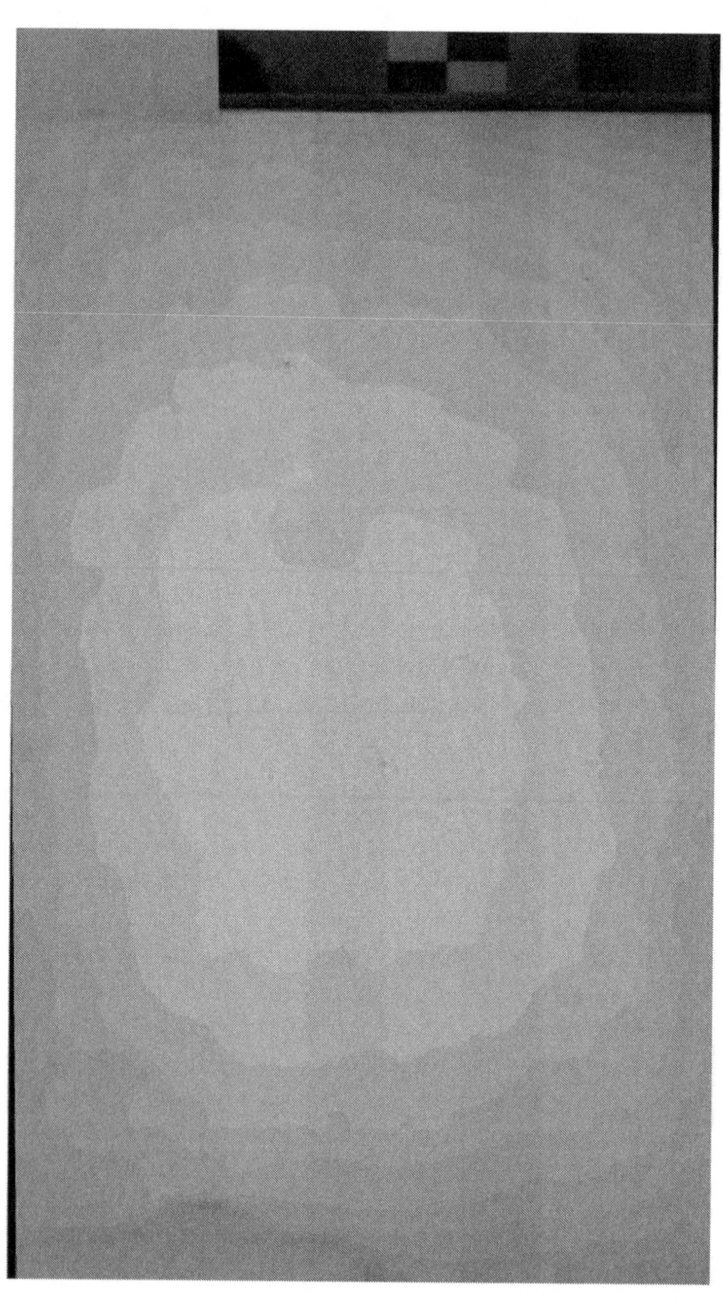

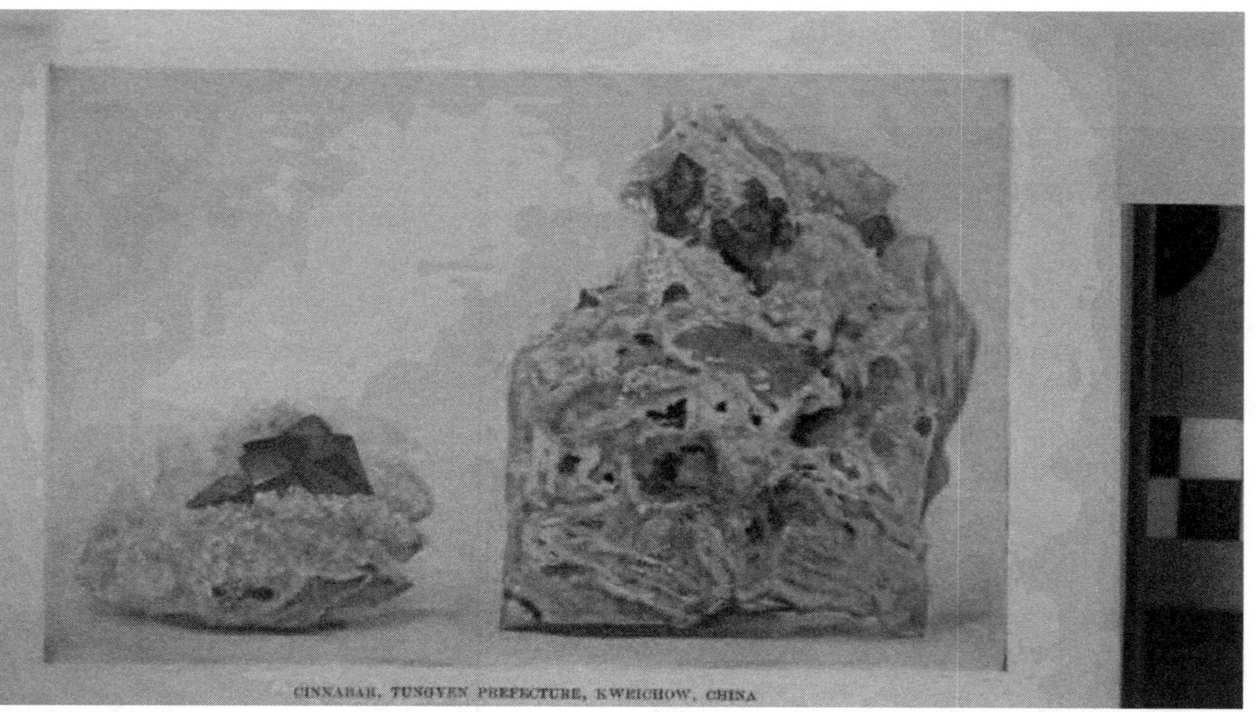

CINNABAR, TUNGYEN PREFECTURE, KWEICHOW, CHINA

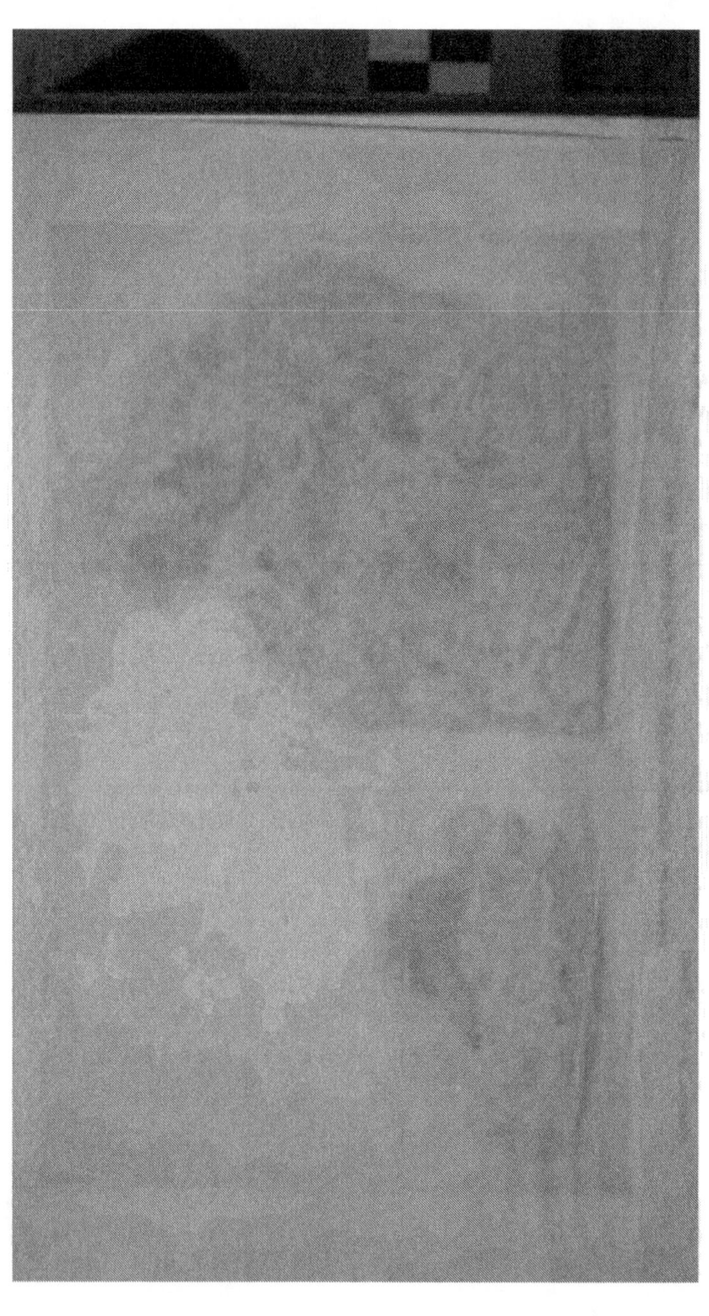

A Book of Precious Stones

THE IDENTIFICATION OF GEMS AND GEM MINERALS, AND AN ACCOUNT OF THEIR SCIENTIFIC, COMMERCIAL, ARTISTIC, AND HISTORICAL ASPECTS

BY

JULIUS WODISKA

WITH 46 ILLUSTRATIONS
IN COLOR AND IN BLACK AND WHITE

SECOND EDITION REVISED

G. P. PUTNAM'S SONS
NEW YORK AND LONDON
The Knickerbocker Press

Seventh Printing
Eighth Printing

Printed in the United States of America

PREFACE

THE object of the author is to gather together in the present volume information of all sorts about precious stones and the minerals which form their bases; it has been his endeavour to include all of the many aspects of his subject, and, at the same time, to present it in such form that it may serve at once as a guide to the professional jeweller, a book of reference to the amateur, and yet prove of equal interest to the general reader.

The study of gems, in its more obvious aspects, forms a division of mineralogy—or more specifically of crystallography—and of the allied science, chemistry; but the author has attempted to avoid the technicalities of these subjects and present the matter in a popular manner.

While it is true that "gemology" may be included under mineralogy or chemistry, nevertheless, so varied are the associations with gems, that if this scientific treatment of them were alone attempted, there would be disregarded some of the most interesting aspects of the subject, which is related not only to art, but to history and even mythology as well.

From all these various standpoints has the subject been approached. The precious stones are described in chapters devoted to each, in the order in which they rank in popular estimation, as are also the more important semiprecious stones, which are classified, those occasionally used being briefly treated. Diamond-cutting, its history and processes, the lapidary and his work, imitations and reconstructed gems, myths and legends, favourite gems of the great, gems and gem minerals in museums, the trade union of the diamond cutters, and the designing and making of jewelry in the new arts and crafts movement, are all considered, and further valuable specific information is comprehended in appended lists, tables, and an extensive bibliography.

In expressing indebtedness to those who have been of assistance to him, the author would first cordially thank his friend Mr. Allen S. Williams, whose scientific knowledge and literary skill have been of very great help in preparing this volume. Among authorities drawn upon for information are Dr. Max Bauer, Professor James

D. Dana, Dr. George Frederick Kunz, Professor Oliver Cummings Farrington, Mr. Edwin W. Streeter, Mr. Gardner F. Williams, Professor Louis P. Gratacap, Mr. Wirt Tassin, and Mr. W. R. Cattelle. Thanks for valuable aid are also due to Mr. Arthur Chamberlain, editor of *The Mineral Collector;* to the editors of *The Jeweller's Circular-Weekly, The Keystone,* and *The National Jeweller;* to The Foote Mineral Company and Mr. William H. Rau, of Philadelphia; to Mr. John Lamont, to Mr. Albert H. Petereit, and Mr. Ludwig Nissen of New York City; to Mr. Walter Scott Perry of Brooklyn, and to the University of California.

The author feels that the experience of more than thirty years as artisan and as manufacturer of jewelry and importer of gems justifies him in presenting in book form the information which is constantly sought of those who are regarded as authorities upon the subject of precious stones.

J. W.

NEW YORK, June, 1909.

CONTENTS

APPENDIX

ILLUSTRATIONS

A Book of Precious Stones

CHAPTER I

GEMS AND JEWELRY—THE INTEREST OF THE SUBJECT, AND THE NEED OF MORE BOOKS CONCERNING IT

FROM the earliest ages jewels have powerfully attracted mankind, and the treatment of precious stones and the precious metals in which they are set, often serves as important evidence, not only concerning the art of early times and peoples, but also concerning their manners and customs. Jewels have been the gifts and ransoms of kings, the causes of devastating wars, of the overthrow of dynasties, of regicides, of notorious thefts, and of innumerable crimes of violence. The known history of some existent famous gems covers more years than the story of some modern nations. Around the flashing Kohinoor and its compeers cluster world-famous legends, not less fascinating to the general reader who loves the strange and romantic, than to the antiquary or the historian or the scientist. These tales of fact or fiction are fascinating in part, because they associate with the gems fair women whose names have become synonymous with whatever is beautiful and beguiling in the sex. In the mind of the lowest savage, as in the thought of man in his highest degree of civilisation, personal adornment has always occupied a prominent place, and for such adornment gems are most prized. The symbolism and sentiment of the precious and semi-precious stones, and precious metals, permeate literature. Jewels have their place in the descriptions of heaven in the sacred writings of almost every people that has attained to a written language.

So wide and so interesting is the subject of precious stones and precious metals, their artistic treatment apart and combined, their importance in society, commerce, and the arts, their part in the wealth of individuals and nations, that it is in a high degree remarkable that, comparatively speaking, so few books have been written about them.

Geology and mineralogy are the names of the sciences that concern themselves with minerals—among them gems—in the rough; metallurgy is the name of the science that has to do with metals; "gemology" is a word sometimes used to describe the branch of art or of the crafts that deals with gems which have passed through the hands of the diamond cutter or the lapidary. The general reader resents the disposition of scientific writers to indulge in technical terminology, though the steady development of popular interest in pure science has in some measure reconciled the reading masses to a sparing and judicious use of the technical terms of specialists.

Scientific hobbies are nowadays common; some take to mineralogy, some to botany, some to entomology. So far as popularity is concerned, the scientific study of gems is, as compared with the studies above named, at a disadvantage. The novice adventuring into the study of nature is apt to be attracted by life and action, and his attention won by the forms that are most beautiful, as birds, butterflies, or wildflowers. Sometimes the adaptability of specimens to photography weighs heavily in the scale of choice, or, perhaps, the ease with which they can be preserved with their natural brilliancy of colouring as in the case of moths, beetles, or the leaves of forest trees. The fascination of penetrating a realm difficult and dreaded, as the reptile kingdom, or of gaining new facts about the life histories of powerful or carnivorous wild beasts proves most potent to some investigators. Geology allures some with its prospecting rambles and the employment found in classifying and installing specimens for exhibition.

The high intrinsic value of diamonds and other precious stones and of precious metals and of all but the least valuable of semi-precious stones, in the rough or in ore, prohibits, for most of us, the possession of representative groups of specimens, and men are not apt to interest themselves deeply in subjects that are difficult of access for the student and observer. This, no doubt, is why the sciences and the arts and crafts immediately concerned with precious stones and their settings can hardly be called popular. Such being the case, there is certainly a place for a book on gems that will be of substantial value to the practical dealer in jewels, to the designer of settings for precious stones, and to the general public who, for a hundred different reasons, are curious in regard to the subjects of which the work treats. It is the author's hope that the present volume will meet the needs of the various classes of readers above referred to, and will at the same time interest them and give them pleasure.

And here the author would lay strong emphasis on one point, namely, that the average jewel merchant or salesman is badly handicapped in his desire to inform himself regarding "gemology," by the lack of reliable and easily accessible books concerned with matters of the first interest to him. There are, to be sure, books, but they are most of them either too technical or too costly. The jewelry trade has its journals, and the best of these offer valuable special information concerning the science and art of gems and jewelry; but, nevertheless, the business man lacks authoritative books which can be understood by readers not possessed of a scientific education. The desire for a special, yet not too technical, literature often finds a voice in the jewellers' trade journals. For instance, in *The National Jeweller and Optician* of April, 1908, there is this complaint: "I know men in the hardware and chemical and other lines who have shelves of interesting books about their lines of commerce right at their hands. This is nowhere the case in our downtown jewelry district. In fact, no trade is poorer in books on the trade than the jewelry and silver and art-metal trades." And in the same issue the complaint is repeated. "It is both astonishing and disappointing," says the journal in question, "that a craft of such antiquity and interest as that of jewelry should have virtually no distinctive literature."

The present volume is designed, as far as it may, to supply the lack alluded to, and to give the salesman and the merchant the kind of information which his customers can fairly expect of him.

CHAPTER II

CLASSIFICATIONS OF PRECIOUS AND SEMI-PRECIOUS STONES

THERE seems to be a considerable difference of opinion among writers on the subject of gems as to those stones which should be classed as precious and those which should be classed as semi-precious. The more scientific writers, from their inclination to treat the matter from the view-point of the mineralogist, appear to be little influenced in their classifications by the inexorable law of demand and supply, or the fickleness of fashion and popular favour. This book, being for the many, will present a classification of the principal gems as handled at the period of its publication by the jewelry trade in America, and classified according to present standards of popularity, or what the authors believe to be such. The arrangement of the scale of popularity is based upon personal experience and observation, and upon the opinions of leading American business concerns engaged in the business of importing and dealing in precious and semi-precious stones, as expressed in replies to letters of inquiry asking for lists of gems classified according to their respective values and the present demand for them. The great divergence of opinion, after the precious stones were set apart, was very interesting. The lists in question were evidently prepared after careful consideration; with most of them there went expressions of doubt as to the propriety or correctness of the arrangement.

Following my nomination of the five precious stones, the semi-precious stones are divided into four classes, the arrangement within each class being alphabetical, because there appears to be no basis upon which it would seem justifiable to give some of these minor gems precedence over others. A number of stones clearly only semi-precious, but which are only occasionally seen by jewellers, are briefly covered in one chapter.

The quintet of gems herein designated as precious stones are accepted as such by all authorities without dissent, with the exception that the pearl is omitted by some devoted scientific mineralogists, because it is not an original mineral. Some writers increase the number of precious stones, as, for instance, Mr. W. R. Cattelle, who includes Oriental cat's-eye, opal, turquoise, alexandrite, and spinel; the last, in trade parlance, being the Balas ruby, and this stone, to the general public, is a ruby.

My classification is as follows:

THE PRECIOUS STONES

Diamond
Emerald
Pearl

Ruby } Corundum
Sapphire }

SEMI-PRECIOUS STONES

Class I

Alexandrite
Amethyst (Siberian)
Aquamarine
Chrysolite (Olivine and Peridot)
Kunzite (Spodumene or Triphane)
Opal (Precious or Noble —of gem quality)
Oriental Cat's-Eye (Cymophane, a variety of Chrysoberyl)
Topaz (Brazilian)
Turquoise

Class II

Beryl
Chrysoberyl
Chrysoprase
Coral
Garnet (Carbuncle, when cut *en cabochon*)
Jade
Tourmaline

CLASS III

Hyacinth
Jacinth
Jargoon
Moonstone
Zircon

CLASS IV

Agate
Amazonite
Aventurine
Azurite
Bloodstone
Labradorite
Lapiz-lazuli
Malachite
Onyx
Sard or Sardonyx

The fact that there is no standard classification of precious stones is curiously illustrated by the great variation exhibited by leading authorities on the subject. Mr. Edwin W. Streeter, the famous English author of books on precious stones, after discussing the various factors of value in several precious stones, writes in the first chapter of his book *Precious Stones and Gems*, as follows:

It is difficult to arrange the various Precious Stones in the order of their relative value, that order being subject to occasional variation according to the caprice of fashion or the rarity of the stones. Nevertheless it is believed that the following scheme, in which all Precious Stones are grouped in five classes, fairly indicates the relative rank which they take at the present day.

I. The Pearl stands pre-eminent. It is true that this substance, being the product of a mollusc or shell-fish, is not strictly a mineral. It is, however, so intimately related in many ways with the family of true precious stones that it properly claims a place in any classification such as that under discussion.

II. In the second class, and therefore at the head of the group of Precious Stones proper, stands beyond all doubt the Ruby.

III. Then comes the Diamond. Many readers may be surprised to find the Diamond taking so subordinate a rank; but the time has gone by when this stone could claim a supreme position in the market. At the present day, the Jagersfontein Mine, in South Africa, produces Diamonds of pure water rivalling the finest stones that were ever brought to light from mines of India or of Brazil.

IV. In the fourth class comes first the Emerald, then the Sapphire, next the Oriental Cat's-Eye, and afterwards the Precious Opal.

V. In the fifth class may be placed such stones as the Alexandrite, the Jacinth, the Oriental Onyx, the Peridot, the Topaz, and the Zircon. Some of these, especially the Alexandrite, are so beautiful that they deserve a more extended use in the arts of jewelry than they enjoy at present.

After these stones comes another class, which may be called the group of *Semi-precious Stones*. Many of these either lack transparency, or possess it in only very limited degree; while those which are pellucid are too common to command more than a trivial value. Such stones are frequently used for inlaid work, or similar ornamental purposes, rather than for personal decoration. As examples of such stones may be cited the Agate, Malachite, and Rock-crystal.

Dr. Max Bauer, in his great work on precious stones, discusses in a very interesting way the motives of mineralogists and jewellers in grouping and classifying gems, and seems to regard each as perfectly justified from their different view-points. As an example he cites the classification by K. E. Kluge, the German authority, as used in his *Handbuch der Edelsteinkunde*, published in 1860, wherein Kluge distinguishes five groups of precious stones, characterised by their value as gems, their hardness, optical characters, and rarity of occurrence. It is interesting to note also that, according to Bauer, Kluge was dominated to a large extent by the then market value of the stones, probably in Germany, or in the European markets in general.

KLUGE'S CLASSIFICATION

1. True Precious Stones or Jewels

Distinguishing characters are: great hardness, fine colour, perfect transparency combined with strong lustre (fire), susceptibility of a fine polish, and rarity of occurrence in specimens suitable for cutting.

A. *Gems of the First Rank*

Hardness, between 8 and 10. Consisting of pure carbon, or pure alumina, or with alumina predominating. Fine specimens of very rare occurrence and of the highest value.

1. Diamond
2. Corundum (ruby, sapphire, etc.)
3. Chrysoberyl
4. Spinel

B. *Gems of the Second Rank*

Hardness, between 7 and 8 (except precious opal). Specific gravity usually over 3. Silica a prominent constituent. In specimens of large size and of fairly frequent occurrence. Value generally less than stones of Group A, but perfect specimens are more highly prized than poorer specimens of Group A.

5. Zircon
6. Beryl (emerald, etc.)
7. Topaz
8. Tourmaline
9. Garnet
10. Precious Opal

C. *Gems of the Third Rank*

These are intermediate in character, between the true gems and the semi-precious stones. Hardness between 6 and 7. Specific gravity usually greater than 2.5. With the exception of turquoise, silica is a prominent constituent of all these stones. Value usually not very great; only fine specimens of a few members of the group (cordierite, chrysolite, turquoise) have any considerable value. Specimens worth cutting of comparatively rare occurrence, others fairly frequent.

11. Cordierite	16. Staurolite
12. Idocrase	17. Andalusite
13. Chrysolite	18. Chiastolite
14. Axinite	19. Epidote
15. Kyanite	20. Turquoise

2. Semi-Precious Stones

These have some or all of the distinguishing characters of precious stones, but to a less marked degree.

D. *Gems of the Fourth Rank*

Hardness, 4–7. Specific gravity 2–3 (with the exception of amber). Colour and lustre are frequently prominent features. Not as a rule perfectly transparent: often translucent, or translucent at the edges only. Wide distribution. Value, as a rule, small.

21. Quartz
    A. Crystallised quartz
        a. Rock-Crystal
        b. Amethyst

c. Common Quartz
d. Prase
e. Aventurine
f. Cat's-Eye
g. Rose-Quartz

B. Chalcedony
a. Chalcedony
b. Agate (with onyx)
c. Carnelian
d. Plasma
e. Heliotrope
f. Jasper
g. Chrysoprase

C. Opal
a. Fire-Opal
b. Semi-Opal
c. Hydrophane
d. Cacholong
e. Jasper-Opal
f. Common-Opal

22. Feldspar
a. Adularia
b. Amazon-Stone

23. Labradorite
24. Obsidian
25. Lapis-lazuli
26. Haüynite
27. Hypersthene
28. Diopside
29. Fluor-spar
30. Amber

E. *Gems of the Fifth Rank*

Hardness and specific gravity very variable. Colour almost always dull. Never transparent. Low degree of lustre. Value very insignificant, and usually dependent upon the work bestowed upon them. These stones, as well as many of the preceding group, are not faceted, but worked by the ordinary lapidary in the large stone-cutting works.

31. Jet
32. Nephrite
33. Serpentine
34. Agalmatolite
35. Steatite
36. Pot-stone
37. Diallage
38. Bronzite

39. Bastite
40. Satin-spar (calcite and aragonite)
41. Marble
42. Satin-spar (gypsum)
43. Alabaster
44. Malachite
45. Iron Pyrites
46. Rhodochrosite
47. Hematite
48. Prehnite
49. Elæolite
50. Natrolite
51. Lava
52. Quartz-breccia
53. Lepidolite

Among the stones enumerated above are some that are never worked as personal ornaments, and many of them have probably never been heard of by American jewellers.

Because of the pre-eminence of Dr. Max Bauer's *Precious Stones*, in the realm which that great work so effectually covers, the arrangement of precious stones made by the distinguished author, and followed throughout in his work, is of interest. It is as follows:

Diamond
Corundum
  Ruby, Sapphire, including star-sapphire and white sapphire, "Oriental aquamarine," "Oriental emerald," "Oriental chrysolite," "Oriental topaz," "Oriental hyacinth," "Oriental amethyst," adamantine-spar.
Spinel
  "Ruby-spinel," "Balas-ruby," "Alamandine-spinel," Rubicelle, Blue-spinel, Ceylonite.
Chrysoberyl

Cymophane ("Oriental cat's-eye"), Alexandrite.

Beryl

Emerald, Aquamarine, "Aquamarine-chrysolite," Golden beryl.

Euclase

Phenakite

Topaz

Zircon

Hyacinth

Garnet Group

Hessonite (Cinnamon stone), Spessartite, Almandine, Pyrope (Bohemian garnet, "Cape ruby," and Rhodolite), Demantoid, Grossularite, Melanite, Topazolite.

Tourmaline

Opal

Precious opal, Fire-opal, Common opal.

Turquoise

Bone-turquoise

Lazulite

Callainite

Olivine

Chrysolite, Peridot.

Cordierite

Idocrase

Axinite

Kyanite

Staurolite

Andalusite

Chiastolite.

Epidote

Piedmontite

Dioptase

Chrysocolla
Garnierite
Sphene
Prehnite
Chlorastolite
Zonochlorite
Thomsonite
Lintonite
Natrolite
Hemimorphite
Calamine
Felspar Group
Amazon-stone, Sun-stone, Moon-stone, Labradorescent feldspar, Labradorite.
Elæolite
Cancrinite
Lapis-lazuli
Haüynite
Sodalite
Obsidian
Moldavite
Pyroxene and Hornblende Group
Hypersthene (with Bronzite, Bastite, Diallage), Diopside, Spodumene (Hiddenite), Rhodonite (and Lepidolite), Nephrite, Jadeite (Chloromelanite).
Quartz.
Crystallised quartz: Rock-crystal, Smoky-quartz, Amethyst, Citrine, Rose-quartz, Prase, Sapphire-quartz, Quartz with enclosures, Cat's-eye, Tiger-eye.
Compact quartz: Hornstone, Chrysoprase, Wood-stone, Jasper, Aventurine.
Chalcedony: Common Chalcedony, Carnel-

ian, Plasma, Heliotrope, Agate with Onyx, etc.

Malachite

Chessylite

Satin-spar (Fibrous Calcite, Aragonite, and Gypsum).

Fluor-spar

Apatite

Iron-pyrites

Hæmatite

Ilmenite

Rutile

Amber

Jet

In an appendix Dr. Bauer places Pearls and Coral.

Of the authorities named as classifying gems, Bauer and Kluge are manifestly moved by their scientific instincts, while Streeter was actuated by popular demand, but responded to temporary conditions and possibly, although maybe unconsciously, to personal interest.

The final test of the rank of gems is their cost in the market, for that tribunal is affected by every factor and influence in the case. The five gems distinguished in this book as "the precious stones" far outclass the gems in the long list that follows in the test of cost, in which all their merits are considered and summed up.

Streeter exalts above all gems the pearl, the mollusc product which Bauer relegates with the comparatively common coral to an appendix. Streeter, who is recognised as a high British authority, accords the ruby second place and places the diamond third; but when he inscribed this judgment "The Syndicate," which now in his own city of London controls with the output of the South African diamond mines the world's gem markets, did not exist. As Streeter was, when he wrote his *Precious Stones and Gems*, expensively and hazardously exploiting the famous ruby mines of Burma, he naturally regarded the ruby as of prime importance.

Kluge's classification is primarily based on the degree of hardness, clearly from the view-point of the strictly scientific mineralogist. Dr. Bauer also yields to the mineralogical influence, for, while he justly leads with the diamond, following it with the ruby and then the sapphire, he continues by naming a line of gems seldom handled, concluding with "Adamantine spar," a name which some jewellers have never heard, nor have they seen the mineral it specifies. This extreme course is pursued by Dr. Bauer because these several stones are alike with the ruby and the sapphire in being the mineral corundum. Dr. Bauer then named spinel, and its varieties chrysoberyl and cymophane, before reaching the noble emerald.

Exceptions may be taken to the order in which semi-precious stones are named by the author by those whose individual experiences in trade have differed; but it is believed that the five precious stones, and the order in which they are named, represent the understanding of American gem dealers and well-informed purchasers, and that the classification of the semi-precious stones fairly represents their general popularity.

Here it may be said, in connection with the influence the value of gems has in their classification, that the price of any kind of precious stone, or of individual specimens, while depending chiefly upon beauty, durability, and similar characteristics, is governed also by extrinsic considerations such as the law of supply and demand and many other things, including fashions, fads, and fancies. A common question propounded to stone merchants is, What is the price of diamonds, sapphires, rubies, or other gems? as though each kind of stone had a common price in the market, like October wheat or steel billets. Each gem stands strictly upon its own merits, and in pronouncing a valuation on it the expert dealer takes into consideration every one of the several factors that are apparent to his keen and reflective examination. Considering the very slight differences involved, or that appear slight to the inexperienced, it is remarkable how nearly several different experts will agree upon the market value of a stone upon which each of them renders an opinion. In the following pages the various precious and semi-precious stones will be considered in the order in which they are arranged in our own classification on pages 9 and 10.

CHAPTER III

THE DIAMOND

THE diamond is generally regarded as the premier gem of the world. Solitary in its chemical composition among precious stones, it is pure carbon, a primary fact that is not as commonly known as it should be and is supposed to be. It seems, indeed, incongruous that such common substances as graphite and lamp-black should be the same, save that they are uncrystallised, as this prince of gems; yet notwithstanding its humble connections, the diamond, in its adamantine lustre, high refraction, reflection, and dispersion of light, and hardness, is alone among minerals. Despite its hardness, the diamond is not indestructible; diamond will cut diamond; it can be burned in the air, being carbon, and will then leave behind carbon dioxide gas and, as ashes, an impurity called carbonado. The facets of a cut diamond can be worn away to some extent by the constant rubbing of fabrics, as is often manifest by contact with wearing apparel. The diamond is also brittle so that it may be easily fractured, especially at the girdle, by striking it a blow against some hard substance, and in a steel mortar with a steel pestle it may be reduced to powder. By what process in Nature's workshop carbon was crystallised into the diamond is unknown, but scientific investigators agree that the process was slow and a prime factor was a titanic pressure.

The specific gravity of the diamond is 3.52; hardness, 10; crystallisation, isometric; cleavage, octahedral and perfect; refraction simple, with an index of 2.439; a high dispersive power; lustre, brilliant adamantine; is combustible though infusible; electric, positively, by friction and a non-conductor of electricity; it is phosphorescent and does not polarise light.

There are three forms of diamonds: crystallised, used as gems; crystalline—imperfect crystallisation,—harder than crystals, termed bort (a word also applied to chips, waste, and stones unfit for cutting); and carbonado, steel gray or black, shapeless, and without cleavage.

To the diamond's surpassing property of dispersing light, or dividing it into coloured rays, is due that fascinating flash of prismatic hues termed its fire. The stone's wondrous brilliancy is due in part to the total reflection of light from its internal faces when the incident ray strikes it at an angle of a little more than twenty-four degrees. Colourless diamonds are richest in the flashing of prismatic hues, while in some coloured specimens it is scarcely apparent; at the same time by-waters, yellow-tinged stones, are sometimes more brilliant in artificial light than are the colourless diamonds.

Diamonds have a wide range of colour; most numerous are the whites, yellows, and browns in a great variety of shades; then come the greens; red stones of strong tints are very rare, as are also blue, which have been fonnd almost exclusively in India; other tints of occasional occurrence are garnet, hyacinth, rose, peach-blossom, lilac, cinnamon, and brown; black, milky, and opalescent diamonds are among the rarities. Diamonds without tint or flaw are rare indeed and even most of the world's famous diamonds have imperfections.

The origin of the diamond's name is the Greek word *adamas*, meaning unconquerable; from the same root spring our words adamant and adamantine.

The origin of the diamond, according to classical mythology, was its formation by Jupiter, who transformed into stone a man, Diamond of Crete, for refusing to forget Jupiter after he had commanded all men to do so.

The diamond is found in alluvial deposits of gravel, sand, or clay, associated with quartz, gold, platinum, zircon, rutile, hematite, ilmenite, chrysoberyl, topaz, corundum, garnet, and other minerals appearing in granitic formations; also in quartzose conglomerates, in peridotite veins, in gneiss, and in eruptive pegmatite.

The ancient source of the world's supply of diamonds was exclusively India; later Borneo produced some, but up to about the year 1700 India was the sole source, and from the anciently famous diamond district and market of Golconda, between Bombay and Madras, in the southern portion, came the Kohinoor, the blue Hope Diamond, and other world-famous gems. The French traveller Tavernier recorded that he visited Golconda in 1665 and that sixty thousand men were employed there; this field is now abandoned. The modern diamond mines of India are in three principal localities. The Madras Presidency in Southern India, which includes the districts of Kadapah, Bellary, Karnul, Kistna, and Godavari, and also ancient Golconda. The second locality is farther north between the Mahanadi and Godavari rivers, and includes Sambalpur and Waigarh eighty miles south-east of Nagpur, as well as portions of Chutia Nagpur province. Bundelkhand, Central India, contains the third region, the principal field being near the city of Panna. The product of all the mines of India has decreased until now it is but a small part of the world's supply.

Borneo's fields produce annually about three thousand carats. The basin of the Kapœas River, on the western slope of the Ratoos Mountain, near the town of Pontianak, is the principal locality.

In 1728 diamonds were discovered in Brazil. They were found by gold miners in river sands, but the finders did not identify the curious crystals sometimes found in their pans when washing the sand for gold-dust and scales. It is related that a monk who had seen diamonds mined in India recognised the characteristics of the Brazilian stones. No sooner had the news of the valuable discovery reached the Portuguese than the King of Portugal seized for the Crown the lands known or thought likely to be diamondiferous. Near Diamantina, in Minas Geraes, the diamonds are obtained from both river and prairie washings. The river deposits are rolled quartz pebbles, mixed with or united by a ferruginous clay of which the usual foundation is talcose clays. Associated minerals include, rutile, hematite, ilmenite, quartz, kyanite, tourmaline, gold, garnet, and zircon. The finest stones result from the prairie washings, where the diamonds occur in a conglomerate of quartz fragments overlaid by earth or sand. Bagagem is a productive locality, and there a fine stone weighing 247½ carats was found. Abatehe, Minas Geraes, is another important field. Diamonds are also found at Lencaes, Bahia; along the river Cacholira, chiefly at Surua and Sinorca, and on the Salobro and other branches of the Pardo River.

The world's diamond markets to-day are almost entirely supplied by the diggings in South Africa, where the discovery of diamonds was so recent as 1867. Children are accredited with the finding of the diamond in South Africa. A Boer farmer, Daniel Jacobs, had a farm near the present town of Barkly West on the Vaal River. On the river's strand were many glittering and coloured pebbles, the only playthings the Jacobs children could get; these pebbles included carnelian, agates and many varieties of

KIMBERLEY, SOUTH AFRICA, DIAMOND MINES. OPEN WORKINGS

quartz, semi-precious stones of some value if cut and marketed in far-off Europe. Among the pebbles which a little son of the Boer farmer brought into the house was a small white stone which sparkled so in the sun that the vrou of the Boer farmer noticed it, although she did not care sufficiently to pick it up, and only mentioned it to a neighbour, Schalk van Niekirk, who asked to see it. The little white pebble had been thrown out, but the children found it in the dust of the yard. Van Niekirk wiped the dust from the stone and found it so interesting that he offered to buy it, which occasioned some mirth, and it was given to him. With a vague instinct that the stone was unusual and had some value, Van Niekirk subsequently asked a travelling trader, John Reilly, to see if he could find out what it was and if anybody would give any money for it. Several merchants in Hopetown and in Colesberg examined it, said it was pretty, and one thought it might be a topaz, but none would give a penny for it. Reilly might have thrown it away but for a casual exhibition of the pebble to Lorenzo Boyes, a Civil Commissioner at Colesberg, who, experimenting, found that the pebble would scratch glass, and seriously said he thought it was a diamond. A local apothecary,

Dr. Kirsh, of Colesberg, hearing the discussion and examining the stone, bet Commissioner Boyes a hat that the stone was only a topaz. The stone was then sent for determination to the leading mineralogist of the Cape Colony, Dr. W. Guybon Atherstone, at Grahamstown, and it was so lightly valued that, to save a higher postage rate, it was mailed to Grahamstown in an unsealed envelope. The expert reported to Mr. Boyes: "I congratulate you on the stone you have sent me. It is a veritable diamond, weighs twenty-one and a quarter carats, and is worth five hundred pounds. It has spoiled all the jewellers' files in Grahamstown, and where that came from there must be lots more. Can I send it to Mr. Southey, Colonial Secretary?"

Upon Dr. Atherstone's report Sir Philip Wodehouse, the Governor at the Cape, bought the rough diamond at Dr. Atherstone's valuation, and the diamond was sent to the Paris Exposition, where it created interest, but no great sensation. Thus a child's find was destined to revolutionise the world's diamond trade, alter the map and the history of South Africa, and place the regulation of the price of the diamond in the hands of a London syndicate.

KAFFIR EMPLOYEES GAMBLING IN THE KIMBERLEY COMPOUND

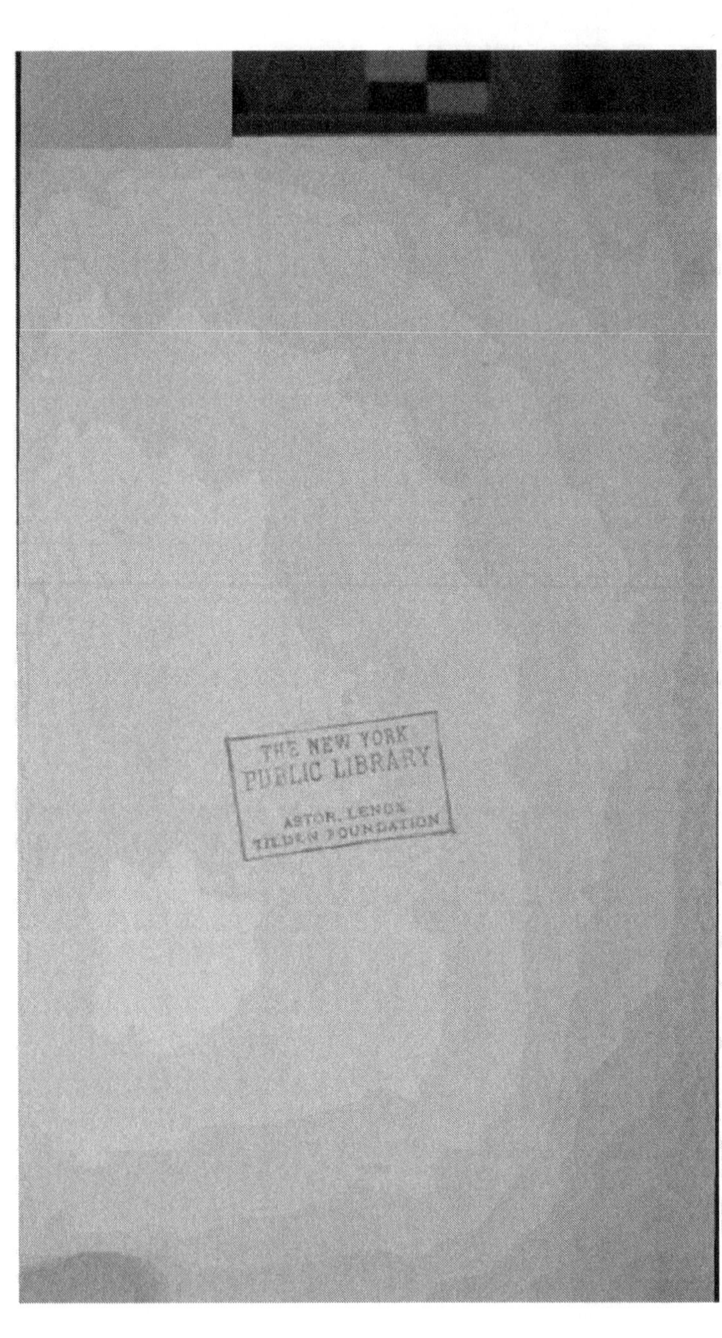

The news of the discovery set Boer farmers in the Vaal valley to some desultory turning over of river gravel in a search for another precious "blinke klippe" (bright stone); but it was ten months before a second diamond was found, and this was on a spot thirty miles away, on the bank of the river below the junction of the Vaal and Orange rivers. In 1868 a few more small diamonds were picked up, and then, in March, 1869, a magnificent white diamond weighing 83.5 carats was picked up by a Griqua shepherd boy on the farm Zendfontein, near the Orange River. Schalk van Niekirk made this poor South African native a local Crœsus by trading for the stone five hundred sheep, ten oxen, and a horse; the thrifty Boer sold the diamond for nearly $55,000 to Lilienfeld Brothers of Hopetown, and Earl Dudley later bought this gem, now the famous "Star of South Africa," for nearly $125,000.

After this, diamond-hunting became more than a pastime in South Africa. The first systematic digging and sifting of the alluvial ground of the Vaal valley was in November, 1869, by an organised party of prospectors from Maritzburg in Natal, initiated by Major Francis of the British Army, then stationed at Maritzburg, and led by Captain Rolleston. The systematic prospecting was begun at Hebron, where the party was joined by two experienced Australian gold diggers named Glenie and King, and also by a trader, named Parker, who, like the Australians, had already been attracted to the locality by the reports of the diamonds found. These prospectors shovelled the river gravel into cradles and pursued the methods of placer washing in vogue in America and Australia. They toiled for many days without sight of a diamond or a grain of gold dust; they then followed the river twenty miles down to Klip-drift, opposite the Mission Station at Pniel; there on January 7, 1870, they found in one of their cradles the first small diamond, the reward of expert methods in the new field. Then came the swarm of diamond hunters.

While the horde of gem seekers toiled and suffered hardships on the Vaal, De Klerk, a Boer overseer on Jagersfontein, the farm of Jacoba Magdalena Cecilia Visser, in a pretty green valley near the settlement of Fauresmith, in the Orange Free State, observed garnets in the course of a little stream, and, having heard that the diggers

...OORS LOADING BLUE EARTH FOR THE WASHING MACHINES, KIMBERLEY MINES

...D PULSATOR, DE BEERS *DIAMOND* MINE AT KIMBERLEY, *SOUTH AFRICA*

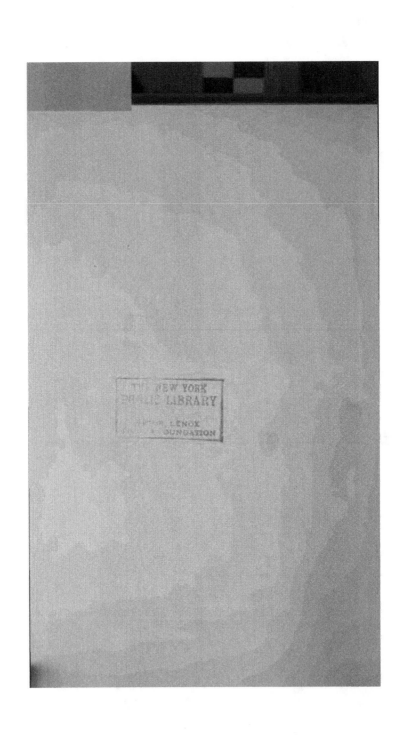

on the Vaal believed the presence of garnets to be an indication of the probable proximity of diamonds, began prospecting one day in August, 1870, and, sifting the gravel in an ordinary wire sieve, at the depth of six feet he found a fine diamond of fifty carats. Soon after, in September, a still more remarkable discovery of diamonds was made at Dutoitspan, on the farm of Dorstfontein, about twenty miles south-east of Pniel; here diamond seeking merged into diamond mining, the diggers penetrating the ground many feet and finding the best stones below the surface. Because of the character of the rotten rock encountered here, the miners made open cuts instead of sinking shafts. The army of diamond seekers spread over the adjoining ground, and early in the year 1871 diamonds were found at Bulfontein, and early in May on De Beers's farm; in July, diamond miners were digging a well for water and, seventy-six feet below the surface, a well-digger was amazed to see a magnificent diamond, which proved afterward to weigh eighty-seven carats, sparkling on the wall of the well. This location was then called—because of the great massing of prospectors there—New Rush or Colesberg

Kopje; this was the beginning of the now world-famous Kimberley mine and the South African mining metropolis of Kimberley.

From this event until 1904, the whole history of South African diamond mining has been ably and thoroughly covered in the copiously illustrated and valuable book of Gardner F. Williams, M.A., entitled *The Diamond Mines of South Africa.* Mr. Williams was long the general manager of the De Beers Consolidated Mines, Ltd., and by experience and known capacity is the recognised authority upon this important subject in the realm of gem history.

A description of the financiering which reconciled warring interests and heterogeneous human elements, to which was added a genius for management which, through science in chemistry, mineralogy, mechanics, and business system, attained the highest degree of economic production and marketing, is not the least fascinating chapter in the wonderful story of the diamond in South Africa. The history of the contest between Briton and Boer, and all else that grew out of the discovery of diamonds on the Vaal, cannot be told here; but the modern methods of extracting the rough diamonds from the blue ground in which they have rested

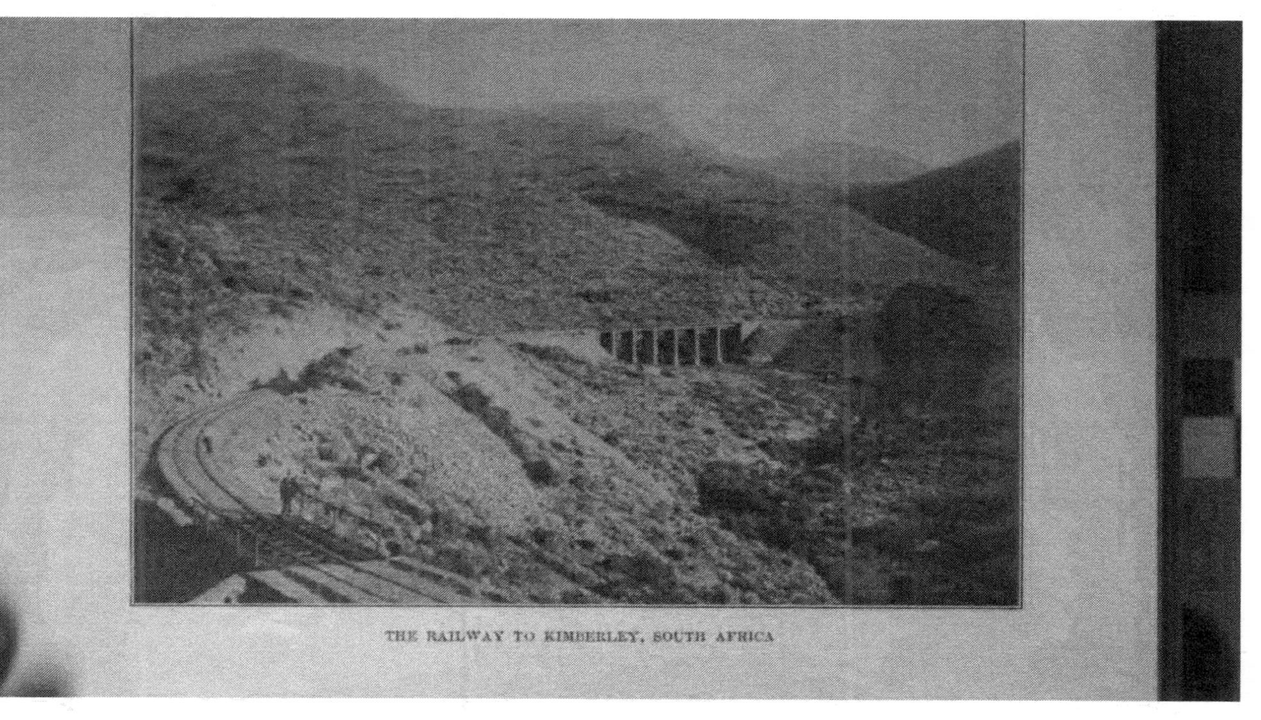

THE RAILWAY TO KIMBERLEY, SOUTH AFRICA

encased for ages is pertinent and worthy of some space in even so compact a book as the present.

The diamond-bearing blue earth from the mines is automatically dumped into ore bins and thence conveyed in trucks drawn by endless wire rope and impelled by steam to the depositing floors on the receiving grounds, which are planed and rolled hard as if for use as tennis courts or brick drying floors. The De Beers mine floors are rectangular sections, six hundred yards long and two hundred yards wide, and extend for four miles; each floor holds about fifty thousand truck loads, a full load weighing about sixteen hundred pounds; spread out until about a foot in thickness, such a load covers about twenty-one square feet. In this great area of blue earth lie the invisible diamonds, for, although some of the rough diamonds may be as large as walnuts, persons walking over the blue earth have almost never seen one. Weathering disintegrates the breccia or blue earth, which process is carried and hastened by wheeled harrows drawn by steam traction engines. Rain accelerates this weathering process and drought retards it. The blue ground from Kimberley mine becomes well pulverised in six months, with the favourable condition of a heavy summer rainfall, while the De Beers earth under similar conditions requires a year's time. About five per cent. of the De Beers mine blue ground is intractable; this, in large pieces, is removed to be reduced by crushers and rolls in the method commonly used for mineral ores. When thoroughly disintegrated the blue ground is hauled to the washing machines to enter the first stage of concentration. Automatic feeders supply the washing machines and the wet mixture from them goes through chutes into a revolving cylinder perforated with holes one and one quarter inches in diameter; lumps too large to pass through these outlets emerge from the ends of the cylinders by way of a pan conveyor to crushing rolls. The pulverised ground which passes through the perforations is fed into shallow circular pans, where the contents are swept around by revolving arms, tipped with wedge-shaped teeth, on a vertical shaft, which forces the diamonds and other heavy minerals to the outer side of the pan, while the thin mud is discharged near the centre through an outlet into which it is guided by an inner rim. The concentrates go from this process into trucks with locked covers in which they are conveyed

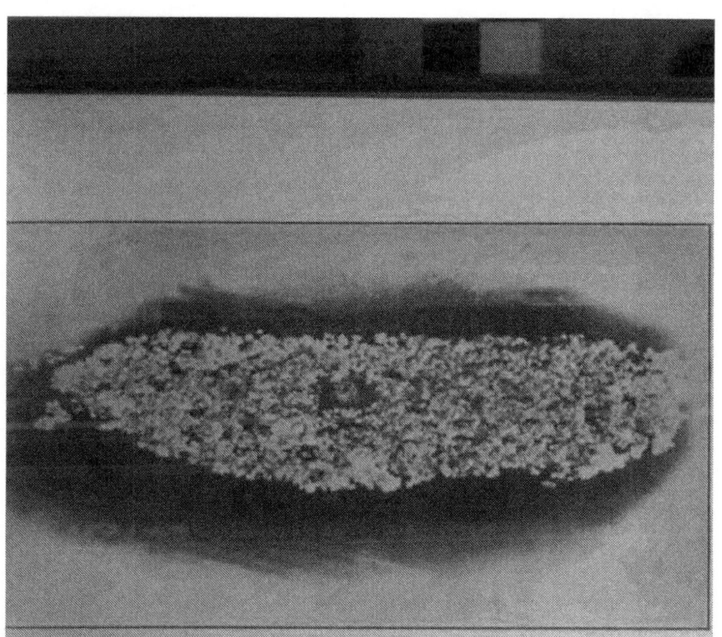

ONE DAY'S DIAMOND WASH AT THE KIMBERLEY MINES

SORTING THE GRAVEL FOR DIAMONDS AT THE KIMBERLEY MINES

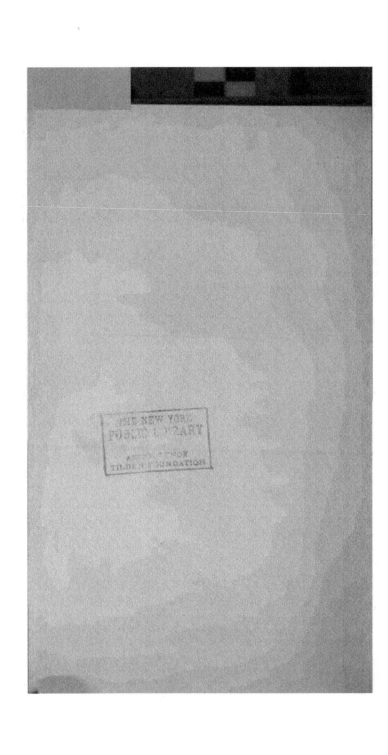

to the pulsator, where they are sifted into five sizes, ranging from one sixth to five eighths of an inch diameter, and passed into a combination of jigs or pulsators with stationary bottoms covered with screens with square meshes a little coarser than the perforated plates of the cylinders that size the concentrate for the jigs. Upon the jig screens, a layer of leaden bullets for the finer sizes and of iron bullets for the coarser sizes is spread, forming a bed that prevents the deposit from passing through the screen too rapidly. The heaviest part of the deposit, with the diamonds, passes through the screens into pointed boxes from which the deposit is drawn off and taken to the sorting tables. The refuse goes to the tailing heap.

But one per cent. of the total amount of blue ground washed goes to the pulsator, and fifty-eight per cent. of this flows over the jigs as waste. Numerous experiments were unsuccessfully made to effect the separation of the diamonds from the worthless concentrates in a less tedious and expensive way than sorting them by hand, when a De Beers employee, Fred Kirsten, suggested coating a shaking or percussion table with grease; and this resulted in the notable discovery that diamonds only, of all the blue ground minerals, adhered to grease, while all else would flow off with water as tailings. The improved shaking tables now used at the South African mines are corrugated, and while a first table fails to detain one third of the diamonds a second table recovers these, almost to the last diamond; so that this invention is practically as certain in its accomplishments as the human eye and hand, while proving a great economy in its operation. It has been demonstrated also that these greased shaking tables will hold other precious stones of high specific gravity and hardness. The diamonds which are heavily coated with grease, of about the consistency of axle grease, by their experience with this process, are cleaned by boiling them in a solution of caustic soda. The quantity of deposit (diamonds) which reaches the sorting tables equals but one cubic foot in 192 cubic feet.

From the sorting tables the diamonds are taken daily to the general office under an armed escort and delivered to the valuators in charge of the diamond department. These experts clean the diamonds of extraneous matter by boiling them in a mixture of nitric and hydrochloric acids, or in fluoric acid. When cleaned

THE TUNNEL ALONG ONE THOUSAND-FOOT LEVEL, DE BEERS DIAMOND MINE, KIMBERLEY, SOUTH AFRICA

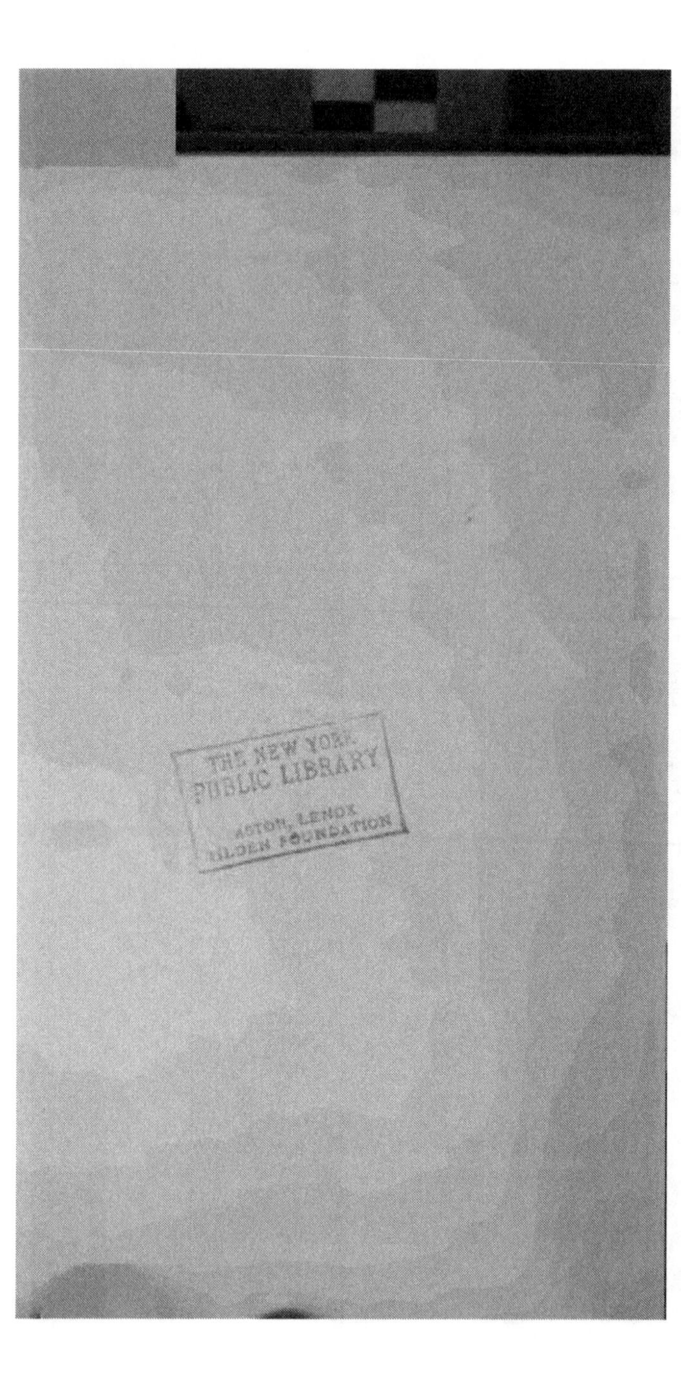

the stones are carefully assorted according to size, colour, and purity, and made up in parcels ready for shipping.

The marketing of diamonds, if fully told, is a story in itself and possesses many phases of interest. Formerly local buyers, who represented the leading diamond merchants of the world, competed at the South African mines for their product, but for the past several years the De Beers Company has sold in advance its annual production to a syndicate of London diamond merchants who have representatives residing in Kimberley, and this is now the medium through which both the product of the De Beers and the Premier mines exclusively reach the markets of the different nations of the world.

The daily production of diamonds is put away in parcels until there has accumulated about fifty thousand carats of De Beers and Kimberley diamonds, the stones from the two sources being mixed, and locally termed "pool goods." The sorters separate and classify them for accurate valuation as follows: 1, Close goods; 2, Spotted stones; 3, Rejection cleavage; 4, Fine cleavage; 5, Light-brown cleavage; 6, Ordinary and rejection cleavage; 7, Flats; 8, Naats;

9, Rubbish; 10, Bort. In the language of the diamond producers "Close goods" are pure stones of desirable shapes; "Spotted stones" are crystals slightly spotted; and "Rejection" stones are those seriously depreciated by spots. "Cleavage" means broken stones. "Flats" are flat crystals formed by the distortion of octahedral crystallisation; and flat triangular crystals—twin stones—are "maacles." The refuse is classed as "rubbish," and common bort or "boart" is polishing material, while round, or shot, bort, found at Kimberley, is now valuable for diamond drill points, since Brazilian carbonado has become scarce.

The first eight classes are further subdivided according to shades, as: Blue White, First Cape, Second Cape, First Bye, Second Bye, Off Colour, Light Yellow and Yellow. Only the "close" or first grade is actually assorted according to these eight shades; with the other grades the sorters are less particular. The ten expert sorters, all Europeans, use no magnifying glasses in their determinations, which are achieved with marvellous accuracy and rapidity. The assorted diamonds are divided into little heaps on a long table covered with white paper; the number of diamonds and their average weights and values are recorded. The buyers for the syndicate of Holborn Viaduct and Hatton Garden, diamond importers of London, pay for their diamonds at the De Beers Company's South African diamond office in cash or bills of exchange on London.

Upon receiving the stones the buyers sort them over to comply with the requirements in London, after which the diamonds, now in from three hundred and fifty to four hundred parcels, each in a specially made paper inscribed with a description of its contents, are packed in tin boxes and these are securely wrapped in cloth-lined packing paper, carefully sealed and delivered to the post-office, which forwards them to Europe as registered mail, the diamonds all being insured during transit in European insurance companies. The syndicate's buyers classify the goods thus shipped as follows: Pure goods, Brown goods, Spotted goods, Flat-shaped goods—all completely formed or crystallised stones; Pure cleavage, Spotted cleavage, Brown cleavage—broken or split stones; Naats or Maacles—flat triangular crystals or twin-stones; Rejections or Bort—diamonds not adapted to or worthy of cutting and used chiefly for splitting and polishing higher grade stones. The higher classes of these are subdivided into six or seven shades and each colour is again subdivided into from eight to twelve sizes.

When the diamonds arrive in London, they are once more reassorted according to the requirements of the trade. The purchasers are dealers in rough diamonds, dealers in brilliants who have their purchases cut and polished for sale, and manufacturers who cut and polish the goods for their own trade, not depending upon the regular diamond-cutting industry.

The selling methods of the famous London Syndicate are peculiar. The different interests present, or represented by experts in the London market, are notified that a "sight" of the goods ready to be disposed of will be afforded on a certain date. The man who contemplates buying for himself or as a representative is compelled by the regulations of this strange market to declare his intentions and to make application to the absolute powers in control of the situation, weeks in advance of the time when a "sight" of the merchandise is expected, for the precious opportunity to buy.

When the favoured business man is admitted to a view of the goods, if he does not buy, he is penalised by being omitted from the purchasing list for six months.

The United States of America is about the only nation that levies a duty on diamonds, under the present tariff, ten per cent. on cut diamonds, while the rough are admitted free. The London Syndicate assorts the diamonds according to qualities, and in general, the American cutters purchase the best. The finest quality, the stones of the purest water, are brought here by American importers and cut in American establishments in a way to satisfy Americans, the most critical buyers of diamonds in the world, who demand the best effects, regardless of waste in diamond-cutting. Even the imported cut goods are frequently recut here.

The other great market for diamonds is Amsterdam in Holland. The industry of cutting diamonds which originated in India, and first appeared in Europe in the town of Bruges—where it was initiated by the Dutch lapidary, Ludwig van Berquen, who invented his particular process in the year 1476—was afterward centred in Antwerp, Belgium. After a struggle for the supremacy, however, Amsterdam became the chief centre of the industry, although it never succeeded in monopolising it, even in Europe. Max Bauer states in his book, completed in 1896, that the diamond-cutting industry in Amsterdam comprised seventy establishments equipped with modern appliances with steam as motive-power; the industry gave employment to twelve thousand persons; one establishment had four hundred and fifty grinding machines and about one thousand employees and in all there were in the diamond city about seven thousand grinding machines (*skaifs*) in operation. American diamond buyers, or jewellers whose interest in that which pertains to their business leads them to visit Amsterdam, the diamond city, while abroad, usually come *via* Cologne. Amsterdam's principal hotel is a rendezvous for diamond importers.

A financial transaction is said to have had much to do with enriching Amsterdam through locating there the centre of the diamond-cutting and polishing industry and making it one of the world's two greatest diamond markets; some rough diamonds deposited in an Amsterdam bank centuries ago as collateral for a loan were ordered, by the bank officials, to be cut. One of the reasons why diamond-cutting as an industry is firmly established in Europe is that there banks make loans on diamonds as collateral.

During the fourteenth century Amsterdam was an asylum for refugee merchants from Brabant; but its enduring prosperity did not begin until the sixteenth century, after the ruination of Antwerp by Spain. The population of Amsterdam, according to a census taken in 1905, was 551,415 and it is now the chief Dutch money market, the home of the Bank of The Netherlands, the diamond-polishing and cutting industry and cobalt blue manufactories being its main industrial interests. The principal square of the city is the "Dam," and canals and well-shaded streets help to make the city picturesque. Places to see in Amsterdam are the Royal Palace, a not particularly impressive building of four stories and painted blue; the "Seaman's Loop," a kind of sailors' club on one side of the "Dam," and the Ryk's Museum, which houses some interesting evidences of Dutch industries as well as much historical material. There are some exhibits of jewelry, gold and silver plate, and art metal work that prove interesting to the visiting foreign jeweller.

But the great feature of the city in the eyes

of the world, its diamond trade, is environed in an unpretentious street about one city block in length, called Tulp Straat; many of the buildings were dwellings now converted into office buildings. The many incongruities here include the existence of a dominant spirit, a species of the genus boss, an untitled ruler of the diamond trade, who is a character worthy a description by Dickens.

A New York diamond merchant at Amsterdam was strolling through the city's streets with this gentleman when he stopped before the bulletin board of a Dutch newspaper and read with great interest some very startling head-lines. The New Yorker waited patiently to hear what the evening edition of an Amsterdam daily newspaper was purveying to its phlegmatic patrons, but the untitled ruler of the diamond trade only said musingly, "Well, you Americans certainly are a great people."

"Why, what have we done now?" asked the American.

"A great people; certainly a great people," reiterated the Hollander.

"Say, what is it?" impatiently demanded the man from Maiden Lane.

"Why, the whole city of Baltimore is burned

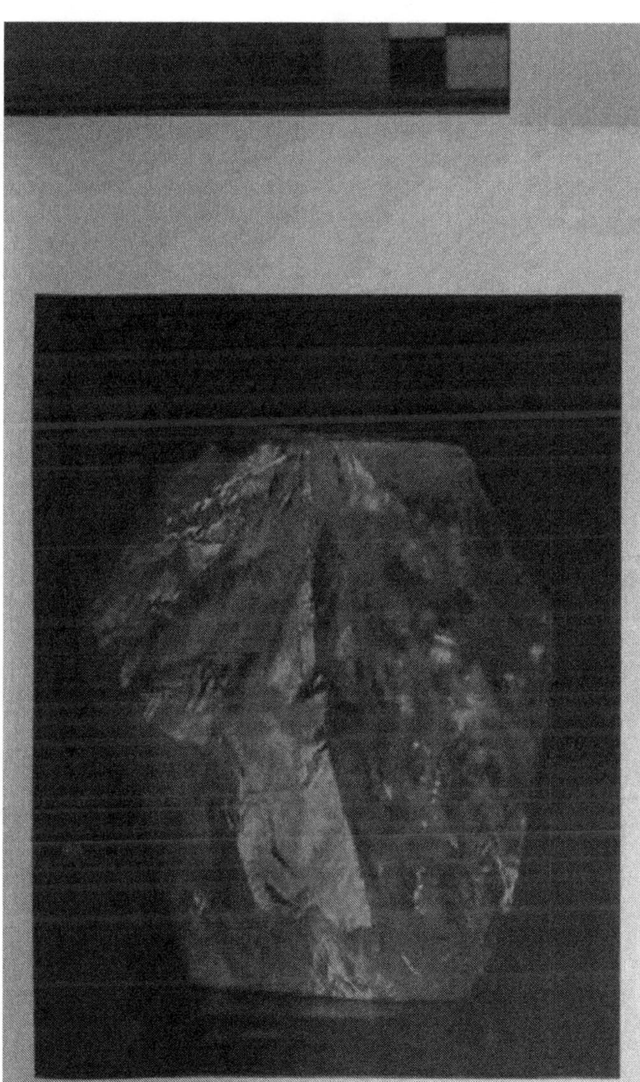

THE GREAT CULLINAN DIAMOND, IN THE ROUGH
Actual size

up; when you Americans do anything you certainly always do it on a large scale," replied the admiring Amsterdammer.

The ways of marketing diamonds to the world are as peculiar in Amsterdam as they are in London. After the diamonds are cut, and polished in the factories by Amsterdam's ten thousand workmen, they are vended through commissioners or through brokers. There is a general meeting ground, a sort of exchange, and there buyers and brokers come together. The space is inadequate and sometimes an overflow meeting of fifty or more men are clamouring for admittance. When they view the merchandise and learn the prices quoted, the buyer who sees something he wants makes an offer; the broker encloses the parcel bid upon in a sealed envelope with the offer made by the buyer written upon it and submits this to owners or persons interested in selling the goods; it is optional for the owner to accept or decline the offer, but if he does accept it, and thereafter the bidder should announce that he had usurped the feminine privilege of changeing his mind, he will find that he must make good his offer or suffer a legal penalty, which might be a term of imprisonment. The diamond brokers of Amsterdam receive a commission from both the seller and buyer.

In Antwerp the principal diamond dealers have their offices in their homes and usually the business is transacted there, or, in some cases, the buyers take the goods with them to their hotels "on memorandum" for leisurely examination before deciding upon their purchases.

The major event of gem history in the year 1908 was the cutting at Amsterdam of the great Cullinan diamond, destined to become the brightest jewel in the British crown. In this connection it may be here mentioned that said crown was already of great weight—thirty-nine ounces and five pennyweights—a handicap that His Majesty King Edward VII. probably does not relish on the rare state occasions when he must submit to having it rest upon his head, as, for example, when it becomes his annual royal duty and prerogative to formally open Parliament. The crown, which usually rests in the Tower of London, contained, prior to additions from the Cullinan Diamond, two thousand eight hundred and eighteen diamonds and two hundred and ninety-seven pearls, besides many other rare and exquisite jewels. Before its eclipse by the Cullinan Diamond, the chief gem ornamenting the crown was a ruby, valued according to an estimate at about $500,000; this famous gem is the one presented to the Black Prince by Spain, in the year 1367, and was worn by Henry V. in his helmet at the battle of Agincourt.

The royal regalia are safely deposited in a chamber of the Wakefield Tower in the Tower of London. The valuable addition resulting from the partitions of the Cullinan Diamond added nothing to the precautions against theft which previously existed. The crown jewels are thoroughly lighted and guarded by night and by day, never, for an instant, being exempt from the scrutiny of armed and uniformed sentries. The jewels are kept in a glass case within a double cage of steel, and cleaned semi-annually under the supervision of high officers of the British realm. The Cullinan Diamonds were on November 1, 1908, delivered to their Majesties, King Edward and Queen Alexandra, at Windsor Castle by Mr. Joseph Asscher of the Amsterdam firm which successfully cut the famous stone. Two secret service men of the Holland government, accompanied by several Scotland Yard detectives, guarded Mr. Asscher's every movement against the possible attacks of thieves. In the following month the Cullinans were conveyed to the Tower by a closely guarded royal messenger in a motor car, and placed with the regalia beside a model of the Kohinoor. Since then the British public and visitors from all parts of the world have curiously viewed the famous gems.

There was disappointment among the diamond cutters and in the gem trade in England when it was decided to send the Cullinan Diamond to Amsterdam to be cut; the great distinction was conferred upon the house of J. Asscher & Co., of Amsterdam and Paris, whose "fabriek," or factory is in the Tulp Straat or "Tol-straat," as it is sometimes written, of Holland's capital. The stone was delivered to the Amsterdam firm in January, 1908, where for nine months it was kept in the vault, of which the walls of concrete and steel are over two feet thick. On February 10th the stone was split by Mr. Joseph Asscher under the supervision of Messrs. M. J. Levy & Nephews, precious stone experts, retained to additionally assure the best scientific methods in the operations in which so vast a sum in values was involved. The stone was first cleft in two

IMPLEMENTS USED IN CLEAVING THE ROUGH CULLINAN DIAMOND

ROOM WHERE THE CULLINAN DIAMOND WAS CUT AND POLISHED, AMSTERDAM, HOLLAND

pieces by Mr. Asscher in such a way that a defective spot in the diamond was exactly in the centre, leaving a part of it on each piece of the stone. Subsequently the larger of these two pieces was split.

The United States consul at Amsterdam, Mr. Henry H. Morgan, forwarded to Washington the best account of the splitting operation that the author has read. After emphasising the delicacy of the work Mr. Morgan described the making of an incision in the stone with a diamond-cutting saw at the point where the stone was to be cleaved and, following the line of cleavage, to a depth of nearly three quarters of an inch. Before the operator were crystal models, cleaved to represent the effect upon the diamond so far as could be indicated in such a manner. In the incision made by the diamond saw a specially made steel knife, comb shaped, without a handle, was inserted; then, while the supervisors and several members of the house of Asscher intently and breathlessly looked on, Mr. Asscher struck the blade on its back with a steel rod and, with the success of the operation still in doubt, all saw the steel knife break against the adamant; again the stroke and with a chorus of sighs of relief the diamond fell in two parts, divided exactly as the expert had planned. The two parts weighed, respectively, 1040½ carats and 1977½ carats. The larger piece was successfully divided late in February, after which the grinding and polishing continued until November. The London *Times* on November 10, 1908, published the first authentic description of the finished Cullinan Diamonds as follows:

In the original state the Cullinan Diamond weighed 3253¾ English carats, or over 1 1/3 pounds avoirdupois. It is now divided as follows: (1) a pendeloque or drop brilliant, weighing 516½ carats, dimensions, 2.322 inches long and 1.791 inches broad; (2) a square brilliant, weighing 309 3/18 carats, 1.771 inches long by 1.594 broad; (3) a pendeloque, weighing 92 carats; (4) a square brilliant, 62 carats; (5) a heart-shaped brilliant, 18⅜ carats; (6) a marquise brilliant, 11¼ carats; (7) a marquise brilliant, 8 9/16 carats; (8) a square brilliant, 6⅝ carats; (9) a pendeloque, 4 9/32 carats; (10) 96 brilliants, weighing 7⅜ carats; and (11) a quantity of unpolished "ends," weighing 9 carats.

The first and second of these stones are by far the largest in existence. Even the second is much bigger than the largest previously known brilliant, viz., the Jubilee, weighing 239 carats, while beside either of them so famous a jewel as the Kohinoor sinks into comparative insignificance, since its weight, 102¾ carats, is little more than one third of that of the smaller, or one fifth that of the larger. Moreover, the stones are not more distinguished for size than for quality. All of them, from the biggest to the smallest, are absolutely without flaw and of the finest extra blue-white colour existing.

As regards the two largest, an innovation was made in the manner of cutting. Normally a brilliant has 58 facets. In view, however, of the immense size of the two largest Cullinan brilliants, it was determined to have an increased number, and to give the first 74 facets and the second 66. This decision has been abundantly vindicated by the results, for the stones exhibit the most marvellous brilliancy that diamonds can show. This fact is all the more remarkable and satisfactory because very large brilliants are apt to be somewhat dull and deficient in fire.

This monumental diamond was found January 27, 1905, on the brink of the open workings of mine No. 2 of the new (Transvaal) Premier mines, near Pretoria, South Africa, by the manager of the mines, Mr. Frederick Wells, an old employee of the Kimberley mines. While making his rounds of inspection Mr. Wells's eye caught a gleam in some debris and, investigating, he perceived that it was undoubtedly a large diamond; placing his find in the pocket of his sack coat he took it to the company's office and its importance was quickly realised. The stone was weighed and found to register exactly 3253¾ carats. Immediately the news was transmitted by telegraph and cable to all parts of the world that the world's greatest diamond had been discovered. The stone was christened "The Cullinan Diamond" after Mr. T. N. Cullinan, the chairman of the Premier (Transvaal) Diamond Company. At the instance of Premier Botha, the Transvaal Assembly presented the great diamond to King Edward VII. in recognition of his granting a constitution to the Transvaal Colony. As stated, the diamond, rough, weighed 3253¾ carats, and measured four by two and one-half by one to two inches. The stone had four cleavage planes, which led experts to surmise that other pieces of the same stone are still in the mines. To one who was not familiar with diamonds the great diamond nearly resembled a piece of ice.

The occurrence of this stone is interesting because it was in a locality that many experts regarded as a place of meagre possibilities, as compared with the steadily producing mines at Kimberley. Diamonds had, indeed, been found in both the alluvial along the Vaal River and in alluvial and in pipes at Rietfontein, near Pretoria. The properties of the Transvaal Mining Company, now the Montrose, were discovered in 1898, as were also those of the Schuller Company; both producing diamonds in profitable quantity, although not comparably with the mines at Kimberley. The Premier (Transvaal) Diamond Mining Company was registered on December 1, 1902, with a capital of £80,000, so that it had been in existence but about two years when it gave the world its record diamond. The Boer War interfered with the development of the mines in the Transvaal. During the year 1899 four companies were registered. After the occupation of the Transvaal by the British, forty-eight companies were registered in the years 1902 and 1903 with an aggregate capital of nearly £2,000,000 sterling.

The new Premier mines are discussed by Mr. Gardner F. Williams in his *The Diamond Mines of South Africa,* in which he expresses doubt that the rich alluvial diggings which resulted from the open works initiated there betokened rich diamond bearing pipes of blue ground. Although the reports of the company showed a large total yield for the number of loads of ground sent to the washing machines, it is pointed out that the ground sent was sorted ground, while that upon which Kimberley statistics are based was not. Mr. Williams stated:

The average value of the diamonds per carat for eleven months was 27s. 4d. The quality of the diamonds in the Pretoria District is poor, the percentage of bort and rubbish being abnormally great. Valued on the same basis, diamonds from the Pretoria District are worth only about fifty-four per cent. of those from De Beers and Kimberley mines.

It is always the unexpected that happens in diamond-seeking. The premises of Mr. Williams and the other experts, who may from personal interest have been subconsciously inclined to make comparisons between Kimberley and Transvaal mines unfavourable to the latter, however sound and scientific, held forth small encouragement to expect great things from the new Premier mines; which, after all, have produced a single gem that outshines anything that the Kimberley mines ever produced.

Until its sun was eclipsed by the revelation of the Cullinan Diamond, the largest diamond which the earth has given to man was the Excelsior, which was ultimately named the Jubilee in honour of the celebration of the sixtieth anniversary of the accession of the late Queen Victoria. The Excelsior-Jubilee was discovered in the Jagersfontein mine in the Orange River Colony, June 30, 1893. The lucky Kaffir who discovered it was rewarded with about $2500 in money, and a horse equipped with a saddle and bridle. The rough stone weighed 971¾ carats, measured two and one-half inches in length, two inches in breadth, and one inch in thickness. Like the Cullinan Diamond, its predecessor had a fault that prevented its becoming a single gem; this was a black spot in the centre which made it necessary to cleave it, as the Cullinan was cleaved. The larger portion was cut into an absolutely perfect brilliant, weighing 239 international carats of 205 milligrams and measuring one and five-eighths inches in length, one and three-eighths in breadth, and one inch in depth. The Excelsior-Jubilee is a blue-white stone of the purest water and in all its qualities approximates perfection. This diamond's predecessor in holding the world's record for weight and size, in the rough, was the "Great Mogul" which is supposed to have weighed 787½ carats. The history of this stone is obscure and so tainted with tradition that the references to it in the various stories of the great diamonds of the world are of doubtful authority.

The romance of gem history is well illustrated by the accepted account of that acme of fine diamond qualities, the Regent or Pitt diamond. Mr. Ludwig Nissen, a New York authority on gems and who talks and writes in an interesting way about them, offers the following narrative as authentic:

The Pitt Diamond, afterward called the "Regent," was found by a slave in the Parteal mines, on the Kistua in India, in the year 1701. The story goes that, to secure his treasure, he cut a hole in the calf of his leg and concealed it, one account says in the wound itself, another in the bandages. As the stone weighed 410 carats before it was cut, the last version of the method of concealment is, no doubt, the correct one. The slave escaped with his property to the coast. Unfortunately for himself, and also for the peace of mind of his confidant, he met an English skipper whom he trusted with his secret. It is said he offered the diamond to the mariner in return for his liberty, which was to be secured by the skipper carrying him to a free country. But it seems probable that he supplemented this with a money condition as well, otherwise the skipper's treatment of the poor creature is as devoid of reason as it is of humanity. The English skipper, professing to accept the slave's proposals, took him on board his ship, and having obtained possession of the gem, flung the slave into the sea. He afterwards sold the diamond to a prominent dealer for a thousand pounds sterling, squandered the money in dissipation, and finally, in a fit of delirium tremens and remorse, hanged himself.

The dealer sold it in February, 1702, to Thomas Pitt, Governor of Fort St. George, and great-grandfather of the illustrious English statesman, William Pitt, for the sum of £20,400. Pitt had the stone cut and polished at a cost of £5000, but the cleavage and dust obtained in the cutting returned to him the handsome sum of £15,000. In 1717 he sold it to the Duke of Orleans, Regent of France, during the minority of King Louis XV., for the sum of £135,000; so that he must have netted a profit of nearly £125,000 on his venture.

Later, in the inventory of the French crown jewels, drawn up in 1791, it was valued at 12,000,000 francs, or $2,400,000. Soon afterwards, during the "Paris Commune," it was, with other valuable jewels, stolen and buried in a ditch to prevent its recovery. One of the robbers, however, on a promise of a full pardon, later revealed its hiding-place, and it was found. All of the criminals were sent to the scaffold, except the one who had turned informer.

The recovery of the "Regent" is claimed to have helped to put the first Napoleon upon the throne of France, by having enabled him, through pledging it to the Dutch government, to raise sufficient funds to make a success of the Marengo campaign. Since its redemption from the Dutch government it has served as an ornament in the pommel of the First Emperor's sword, and has ever been the most conspicuous gem of the crown jewels of France. It now quietly rests to meet the wondering eyes of the world's tourists in the Galerie d'Apollon in the Louvre, Paris.

Though a rich and valuable treasure, the "Pitt" or "Regent" has unquestionably been the cause of more misery than joy. It sent the first dishonest holder to a watery grave, the second to the rope, and the third, which consisted of several, to the guillotine; though it also restored the fortunes of an ancient English family, which subsequently gave to England her most distinguished statesman, and is said to have helped in the creation of an empire and in the making of one of the world's most famous characters.

The most recently discovered diamond field that holds forth promise of an output sufficient to affect the world's market for diamonds is in Germany's colonial possessions in southwest Africa, and if it results in great wealth for the Fatherland it will be warmly welcomed as a compensation in part for the millions that Germany's exploitation of the region has cost, chiefly because of intractable warring natives. The new field is near Lüderitz Bay, and a remarkable feature is that the diamonds are found separately in a coarse sand. Twelve of the best stones among the first found were sent as a gift to Emperor William by his loyal subjects, the colonists. Never before was the marketing of precious stones so carefully planned in advance of their production. The output will be strictly limited, following the policy of the English Syndicate, and the mining will be closely regulated by the German government. The annual product is expected to reach about 140,000 carats. The syndicate is reported to be composed of representatives of leading German banks and various combinations of speculative investors in diamond corporation shares; among them are the Lenz-Stauch-Nissien group, the Berlin Commercial Co., and Kohnanskop group. The last is of minor importance and is controlled by Englishmen. It is agreed that all stones are to be sent to Lüderitz Bay, where they will be taken by the syndicate. The companies that deliver will receive at once a part payment to cover cost of mining. The stones will be weighed, packed, and sent to Berlin under the owners' names, where they will be sorted and sold and owners credited with the profit.

No definite arrangements have been made to establish a German diamond market. It seems improbable that either Hanau or Frankfort will be considered. Berlin seems to meet all the vertically. The specific gravity of the transparent flawless beryl is 2.73, usually 2.69 to 2.70; hardness, 7.5 to 8; brittle; cleavage indistinct; fracture uneven to conchoidal; lustre vitreous, sometimes resinous. Beryl colours include emerald green to pale green, pale blue, pale yellow, honey, wine and citrine yellow, white, and pale rose-red. Pleochrism is unusually distinct, sometimes strong, in the emerald especially, which through the dichroiscope reveals two different shades of green.

Beryl includes the emerald, aquamarine, goshenite, and davidsonite. The differences are principally in colour.

Beryl is a silicate of the metals aluminium and beryllium, containing the oxide alumina in small amount, which is, however, a more important constituent in corundum, spinel, and chrysoberyl. There is some variation in beryl from different localities; the chemist Lewy, who analysed the beautiful emerald beryl that is found at Muzo in Colombia, South America, found: silica, 67.85; alumina, 17.95; beryllia, 12.4; magnesia, 0.9; soda, .07; water, 1.66; and organic matter 0.12, besides a trace of chromic oxide. An analysis of a specimen of aquamarine from Adun-Chalon in Siberia by Penfield

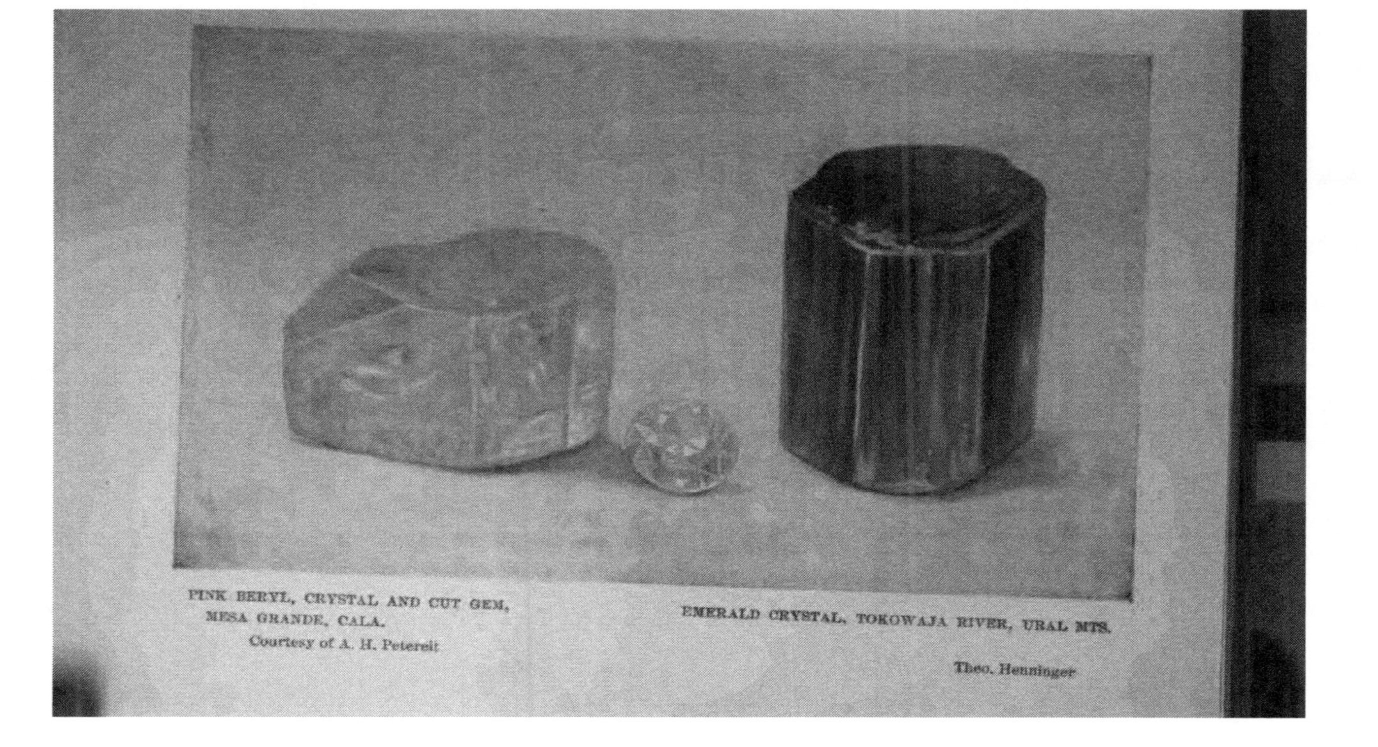

PINK BERYL, CRYSTAL AND CUT GEM, MESA GRANDE, CALA.
Courtesy of A. H. Petereit

EMERALD CRYSTAL, TOKOWAJA RIVER, URAL MTS.

Theo. Henninger

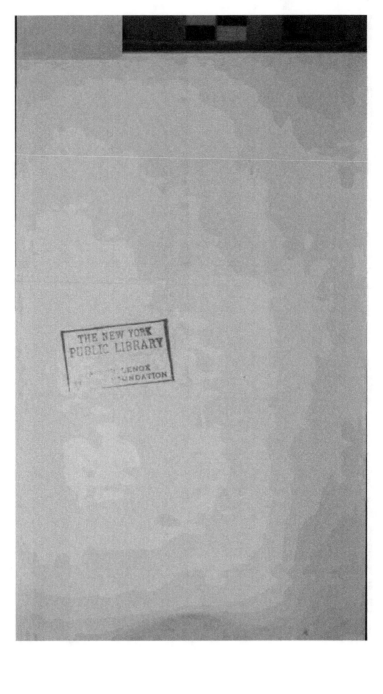

resulted in: silica, 66.17; alumina, 20.39; beryllia, 11.50; ferrous oxide, 0.69; soda, 0.24; water, 1.14, and a trace of lithia.

The only acid which will attack beryl, so far as has been discovered, is hydrofluoric acid. Before the blowpipe beryl becomes white, cloudy, and fuses, but only with difficulty, at the edges to a white blebby glass.

Beryl, like all other hexagonal crystals, is bi-refringent, but only to a small extent. The beauty of beryl, therefore, depends not upon a play of prismatic colours, but upon unusually strong lustre and a fine body-colour. The bright grass-green beryl is the emerald; the pale varieties are styled precious or noble beryl. Aquamarine is pale-blue, bluish-green, or yellowish-blue; the yellowish-green variety is called aquamarine-chrysolite; jewellers call the yellow variety beryl and the pure golden-yellow golden beryl. The dichroism of all transparent varieties of beryl can often be discerned with the eye unaided by the dichroiscope; this property usually suffices to clearly distinguish beryl from any imitations. A curious characteristic of the emerald beryl is that its colour is by no means always uniformly distributed through the body of the stone; the different coloured portions may occur in layers or irregularly; when in layers the layers are usually perpendicular to the faces of the prisms.

The high esteem in which choice emeralds are held and the high cost of this gem are due in great part to the rarity with which a gem approximating perfection occurs. Most of the grass-green beryl crystals are cloudy and dull; these disqualifications are due to fissures and cracks, but also to infinitesimally small enclosures of foreign matter, either fluid or solid, such as scales of mica. When clouded by fissures emeralds are called by jewellers "mossy."

A "perfect" (approximately of course) emerald-beryl stone is worth nearly, sometimes fully, as much as a fine natural ruby and more than a diamond—that is, a stone of one carat or thereabouts,—while large stones are so rare that they bring fancy prices out of all proportion to their size. The average emerald beryl fit for cutting is but a small stone. Tradition and unscientific accounts tell of phenomenally large emeralds, but one of the largest and finest actually known to exist belongs to the Duke of Devonshire; this is a natural crystal, measuring two inches across the basal plane, and weighs 8 9/10 ounces, or 1350 carats; in colour, transparency, and structure it is almost without a fault. This fine stone was found in the emerald mines at Muzo in Colombia, South America. Another large crystal known belongs to the Czar of Russia; its measurements are reported to be twenty-five centimetres (nearly ten inches) in length and twelve centimetres in diameter.

The character of each piece of the rough beryl placed in the hands of the lapidary decides what cut shall be applied to an emerald. Small stones are usually cut as brilliants or rosettes, while the large ones are sometimes cut as a simple table stone, or more generally step-cut with brilliant facets on the upper portion. Cut gems of good colour and transparency are mounted in an open setting; paler stones were formerly, in Europe, reinforced with a green foil beneath them, while fissured or faulty stones were mounted in an encased setting with the bottom blackened. As natural crystals of beryl are large the gems are often extracted from the mass by expert and skilful artisans who saw the crystals into the desirable sizes.

The emerald beryl might be truly said to be a precious stone of strong individuality, for, besides its characteristic of an uneven and irregular distribution of colour, it is unique geologically, for it occurs exclusively in its primary situation, that is, in the rock in which it was formed. It is one of the minerals characteristic of crystalline schists, and is frequently found embedded in mica schists and similar rocks. The magnificent beryls found at Muzo, Colombia, however, are an exception; there the emeralds are embedded in calcite veins in limestone. Emeralds are never found in gem gravels, like diamonds, rubies, sapphires, and other precious stones.

The ancient source of the emerald was Ethiopia, but the locality is unknown. From upper Egypt, near the coast of the Red Sea and south of Kosseir, came the first emeralds of historic commerce. There is a supposition that the emerald beryl was first introduced commercially into Europe just prior to the seventeenth century from South America. Emeralds had been found before this, however, in the wrappings of Egyptian mummies and in the ruins of Pompeii and Herculaneum. Ancient Egyptian emerald mines on the west coast of the Red Sea were rediscovered about 1820 by a French explorer, Cailliaud, on an expedition organised by

Mehemet Ali Pasha; the implements found there date back to the time of Sesostris (1650 B.C.). Ancient inscriptions tell that Greek miners were employed there in the reign of Alexander the Great; emeralds presented to Cleopatra, and bearing an engraved portrait of the beautiful Egyptian queen, are assumed to have been taken from these mines. Cailliaud, under permission of Mehemet Ali, reopened the mines, employing Albanian miners, but, it is supposed because only stones of a poor quality were found, the work was soon and suddenly given up.

The Spanish *conquistadores* found magnificent emeralds in the treasure of both Peru and Mexico, but none are now found in those countries. An immense quantity of emeralds, many of them magnificent, and a large proportion of which are probably still in existence in Europe, was sent to Spain from Perù. The only place in the new world that the Spanish found emeralds by prospecting for them in the earth, was in Colombia or New Granada; perhaps the gems of the Aztec sovereigns and the Incas came from there.

The Spaniards first learned of the existence of the Colombian emeralds on March 3, 1537, through a gift of emeralds by the Indians, who, at the same time, pointed out the locality from which they were taken; this spot, Somondoco, is now being mined by an English corporation, although only second-class stones have been found there by these modern emerald miners. Muzo, where the present supply of the world's finest emeralds is mined, is about one hundred miles distant in the eastern Cordilleras of the Andes on the east side of the Rio Magdalena in its northward course. The only other locality of importance where emerald beryls are now found is about fifty miles east of Ekaterinburg in the Ural Mountains, Siberia, where Uralian chrysoberyl, or alexandrite, is found. The grass-green beryl is also found in an almost inaccessible locality in the Salzburg Alps.

Fine emeralds have been found in the United States, the most notable locality at Stony Point in Alexander County, North Carolina, but the supply at this place seems to be exhausted.

The name "emerald" applied indiscriminately to green transparent, translucent, and even opaque stones, complicates, to the inexpert, everything about the emerald question; for instance, it was long assumed that emeralds came from Brazil and green stones were called "Brazilian emeralds." There is no authentic proof that a true emerald was ever found in Brazil, and it is supposed that green tourmalines found there account for the "Brazilian emerald" myth. In ancient times the name emerald was applied to green jasper, chrysocolla, malachite, and other green minerals. There is still a custom of calling stones other than beryl "emerald," with an explanatory prefix. Thus, Oriental emerald is green corundum; "lithia emerald" is hiddenite, a green mineral of the pyroxene group occurring associated with the emerald beryl in North Carolina. "Emerald-copper" is dioptase, the beautiful green silicate of copper. Among the green minerals sometimes sold under the name of emerald are: the green corundum, demantoids, or green garnets, hiddenite, diopside, alexandrite, green tourmaline, and sometimes chrysolite and dioptase. These minerals are all of higher specific gravity than beryl and all can be distinguished from beryl emeralds by tests possible to the scientific gem expert.

CHAPTER V

THE PEARL

IN its purity, liquid beauty, and charm of romantic and poetical association the pearl—aristocrat of gems—leads even its peers of the highest rank, the diamond, emerald, ruby, and sapphire. The sea-gem has throughout all recorded time formed the fitting necklace of feminine royalty and famous beauty; the state decorations of dusky Oriental potentates and their principal treasures have been pearls. From the ocean's bed and the turgid streams of midland North America, from almost anywhere that is the habitat of the oyster or the humble mussel come these pale, lustrous treasures that may prove to be almost priceless. The existence and recognition of the beauty of the pearl as a personal ornament and treasure is undoubtedly prehistoric on every continent. The discoverers and *conquistadores* from old Spain found quantities of them in the western Indies, on the Spanish Main, in Florida, Mexico, and

Peru; the mound-builders of North America possessed them; in the far East they were cherished centuries before the then Western world of Europe knew them; there is said to be a word meaning a pearl in a Chinese dictionary four thousand years old, and who knows how old is their presence in India.

Pearls were in the jewel caskets of Egypt's Ptolemies; and the first jewel mentioned in the most ancient decipherable and translatable writings extant is the pearl, and its identity is unquestioned, because the gem of the sea is solitary among jewels and is not to be confounded with the hard mineral gems which, even to-day, with all the advance in scientific knowledge, are constantly becoming mixed in the minds of men. From written records the modern ken of pearls extends back about twenty-three hundred years, and we hear of them in the writings of Pliny, the indefatigable investigator and disseminator of what he believed to be facts about almost everything in nature, who four hundred years later gathered together the knowledge of his day about pearls and included it in his voluminous literary grist.

In the technical literature of the United States National Museum, the pearl is coldly and remorselessly comprehended under the generic term "carbonate of lime" along with the beautiful but less valued coral, which is also a product of the sea; and marble, which concerns architects and sculptors, more than gem fanciers; and calcite and aragonite, which are varieties of satin spar and far down in the gem stone scale of hardness. It seems almost like desecration to reduce the lustrous pearl of peerless beauty and royal and romantic associations to the concrete mineralogical base of carbonate of lime; but thus are the insistent requirements of the mineralogists conserved. Therefore, pearls are concretions of carbonate of lime found in the shells of certain species of molluscs. An irritation of the animal's mantle promotes an abnormal secretory process, the cause of the irritation being the introduction into the shell of some minute foreign substance, sometimes a grain of sand.

The lustre of pearls is nacreous, which means resembling mother-of-pearl, a lustre due to the minute undulations of the edges of alternate layers of carbonate of lime and membrane. The lustre of some pearls exists only on the surface; the outer surface of others may be dull and the inner lustrous. The specific gravity of the pearl is 2.5 to 2.7; hardness, 2.5 to 3.5. The shape varies and the range of size and weight is great. The smallest pearl in commerce is less than the head of a pin; the largest pearl known is in the Beresford Hope collection in the Museum at South Kensington, London. Its length is two inches and circumference four and a half inches. It weighs three ounces (1818 grains).

Although the whiteness of the pearl is constantly used for comparison, pearls range in colour from an opaque white through pink, yellow, salmon, fawn, purple, red, green, brown, blue, black, and in fact every colour and several shades of each; some pearls are also iridescent. The colour and lustre are generally that of the interior shell surface against which the pearl was formed.

The beauty and value of the pearl, in brief, depend upon colour, texture, or "skin" transparency or "water," lustre, and form; pearls most desired are round or pear-shaped, without blemish, and having the highest degree of lustre. The queen of existing pearls is La Pellegrina now in the Museum of Zosima, Moscow, Russia. La Pellegrina is perfectly round and of an unrivalled lustre. It weighs 112 grains.

While individual pearls or strands of them may be worth a prince's ransom, their beauty and value are not immutable; pearls may deteriorate with age or be sullied by the action of gases, vapours, or acids, and the known methods for their restoration to their original appearance and value are not always successful. Fine pearls should be carefully wiped with a clean soft cloth after they have been worn or exposed, and kept wrapped in a similar fabric in a tightly closed casket.

Pearls are found in nearly all bivalves with nacreous shells, but the principal supply is derived from a comparatively few families, led by the Aviculidæ, Unionidæ, and Mytilidæ. The first group includes the pearl oyster of the Indian and Pacific oceans, from which has come the bulk of the world's pearls; the second includes the unio, or fresh-water mussel of North America; and the third is a family of conchiferous molluscs, mostly marine, the typical gems being *Mytilius edulis,* or true mussel, which has a wedge-shaped cell and moors itself to piles and stones by a strong coarse byssus of flaxy or silky-looking fibres. The distribution of these molluscs is world-wide.

"In all ages, pearls have been the social

insignia of rank among the highly civilised," writes W. R. Cattelle in his standard book *The Pearl*. First lavishly used by the princes of the East for the adornment of their royal persons, as the course of empire trended westward the pearl followed the flag of the conquerors, and thus, in time, as Rome's power and affluence grew into world-control, her treasure of pearls grew to vast proportions and became identified with the social eminence and arrogance of the Cæsars and patrician Rome. To-day the market for the best in pearls of recent finding, as for all new products of precious stones, or for famous jewels, whose owners' changing fortunes bring them to the parting, is within the new régime of Crœsus represented by the multi-millionaires of the United States. The world's best buyers of jewels are not always as willing to have their princely expenditures known as is generally believed, and the names of some of America's heaviest purchasers of gems have not been revealed by the dealers. It is authoritatively stated that the finest single strand of large pearls in existence was recently acquired by a Western millionaire of the United States. The strand is composed of thirty-seven pearls ranging from eighteen to fifty-two and three-

quarter grains each, the latter being the largest central pearl. The pearls combined weigh 979¾ grains, and the strand is said to have cost its possessor $400,000.

CHAPTER VI

RUBIES

ALTHOUGH we place the ruby fourth among the precious stones, so few are the superior rubies in commerce, or that the world sees, that when a perfect ruby of the weight of ten or more carats enters the market, it brings a price three times as great as does a diamond of the same weight.

The natives of India indiscriminately apply the name "ruby" to all coloured precious stones, and it is the habit of American dealers in precious stones to be almost as general in calling various red gems rubies, although they do distinguish by calling the corundum ruby "Oriental ruby." This being a book for everyone, other red stones commonly or even occasionally appearing in the jewelry trade and called by merchants rubies will be comprehended and described in this chapter, leading with the corundum reality, which is beyond compare.

Corundum crystallises in the hexagonal system in six-sided prisms and pyramids, the crystals frequently being rough and rounded; hardness 9; brittle; specific gravity 3.9 and upwards to 4.16; lustre adamantine to vitreous; sometimes the lustre is pearly on the basal plane; and occasionally there is exhibited a bright, opalescent, six-rayed star in the direction of the vertical axis. The colour range is almost unlimited, blue corundum being sapphire. The strongly coloured varieties are pleochroic. Corundum is sometimes phosphorescent, with a rich red colour. The red-coloured corundum or ruby varies from a rose to a deep carmine, the desideratum being a "pigeon's blood" red, and the same crystal will sometimes reveal different colours. Like its brother in the noble corundum family, the ruby is a peer of the realm of precious stones, and second only to the throne of the sovereign diamond.

In chemistry, corundum is pure alumina, the oxide of the metal aluminum, composed of 53.2 per cent. of the metal and 46.8 per cent. of oxygen. Natural corundum is probably never chemically pure; the inclusions of foreign elements, sometimes but the merest traces, impart the colour that makes the gem. When foreign matter is present in large proportion corundum is impossible for gem purposes, although of great value industrially; inferior translucent specimens serve for pivot supports of watches and other delicate machines and the opaque as an abrasive; thus common corundum is used for cutting and polishing gem minerals lower in the scale of hardness than the diamond, a variety of it being the common compact black emery powder used for sharpening and polishing in mechanical and domestic uses, and familiar to everyone.

A chemical analysis of a fine specimen of an "Oriental ruby," of the approved rich deep red hue was as follows: alumina, 97.32; iron oxide, 1.09; silica, 1.21; in all, 99.62. The extent to which crystallography goes and its fine, yet plain, distinctions, in determining gem minerals, are illustrated by the marked crystallographic differences between the ruby and the sapphire, which differ but slightly in chemical composition, having the same constituents but different proportions; thus one typical sapphire analysed entire exhibited alumina, 97.51; iron oxide, 1.89; and silica, 0.80; in all, 100.20. The forms of corundum generally occur in two different habits represented by the ruby and the sapphire; in the former the prism predominates and in the latter the hexagonal pyramid.

Although corundum is second to the diamond in point of hardness, it is approached much more closely by the minerals next below it in the scale of hardness, than it approaches the eminent and reserved diamond.

Pure corundum has a high specific gravity ranging from 3.94 to 4.08, and this great density makes the specific gravity test in distinguishing it from other stones both easy and important. The differently coloured varieties have not been proved to vary in this particular. Acids will not attack corundum nor is it fusible before the blowpipe. Some specimens when heated in the dark are beautifully phosphorescent. Corundum, by friction, develops positive electricity, which it retains for some time. The lustre of corundum and its fire approach these qualities in the diamond, but the lustre is vitreous instead of adamantine, although it is very durable. Corundum is optically uniaxial and strongly doubly refracting, but the dispersion produced is slight and it is, therefore, incapable of emitting flashes of prismatic colours like the diamond. Coloured corundum crystals are dichroic and the deeper the colour the more pronounced the dichroism. A constant characteristic of coloured corundum gems is that they are as beautiful by artificial light as by daylight.

There are at least nine varieties of corundum used as gems and familiar to nearly all jewellers; the coloured varieties, other than the red ruby and blue sapphire, are named for the gems of other mineral species that they resemble in colour, only with the distinguishing prefix of "Oriental." The arbitrary names and colours are: Ruby ("Oriental ruby"), red; Sapphire ("Oriental sapphire"), blue; Leuco-sapphire (White sapphire), colourless; "Oriental aquamarine," light bluish-green; "Oriental emerald," green; "Oriental chrysolite," yellowish-green; "Oriental topaz," yellow; "Oriental hyacinth," aurora red; "Oriental amethyst," violet.

The colour-varieties of corundum are found in irregular grains and as crystals embedded in some old crystalline rock, as granite or gneiss. The gem-varieties frequently occur as secondary contact minerals, which contact with a molten igneous rock has developed in limestone. These embedded crystals are frequently liberated by the weathering and uncovering of such rocks, and then the crystals are found in the débris in the beds of streams.

Red corundum is supposed to be identical with the anthrax mentioned by Theophrastus and to have been termed carbuncle during the Middle Ages. The colour-tone of the ruby varies greatly, and the presence of deep, intense tones of red causes the term "masculine" to be applied to a gem, while the paler tints suggest the term "feminine." Rubies range from a delicate pink tint through pale rose red to reddish-white, pure red, carmine red, or blood red. A tinge of blue or violet is frequently discernible in these shades. The desired tone in ruby colour was so aptly compared by the Burmese to the blood of a freshly-killed pigeon that the term "pigeon-blood" is the accepted qualification for the colour of the choicest and costliest ruby gems. The colouring is not always uniform, there sometimes occurring alternate layers of colours and colourless stone; a process of heating usually renders the colour uniform. The ruby does not lose its colour when heated, and hence it is assumed that the colouring matter is not organic, as in that case it would be destroyed, but is probably due to a trace of chromium. A graduated increase of heat will not fracture the stone, which upon cooling becomes white, then green, and finally regains its original red colour. The ruby is dichroic according to the direction in which it is seen, and in cutting it this must be taken into consideration; the table—the largest facet surface—should be aligned with the basal planes of the crystal, in order to exhibit the greatest possible depth of colour. The dichroism of the ruby is one of its certain distinctions from spinel, garnet, and other red stones which crystallise in the cubic system and therefore are but singly refracting.

Rubies sometimes show on their basal planes, or on a convex surface which corresponds to the bases, a six-rayed star of gleaming light; these are called asteriated rubies, "star-rubies," or ruby cat's-eye.

So valuable are flawless rubies of good colour, that when they ascend much above a carat in weight their prices depend to a considerable extent on fancy. A three-carat ruby of desirable qualities is a rarity, while three-carat diamonds are common. Although nothing will definitely indicate what a fine ruby of three carats and upward might bring in the open market, yet Dr. George F. Kunz appraised a fine ruby of

9 5-16 carats at $33,000, and Mr. E. W. Streeter, the London jeweller and author, records a purchase price of about $50,000 for a cut ruby of 32 5/16 carats.

The common faults of rubies are lack of clearness; the presence of "clouds," also termed silk, especially in light-coloured stones; patches which resemble milk ("chalcedony patches"); internal cracks and fissures ("feathers"); and the colour being unequally distributed.

From the beginning of its history the main supply of the beautiful ruby gem has been from a small territory in upper Burma, whence, also, have come those of the finest quality. The centre of this mining region and the ruby trade is the town of Mogok, ninety miles north-north-east of Mandalay. The mining district ranges from four thousand to nearly eight thousand feet above sea-level, but, despite its altitude, this forest-covered region proves unhealthy for Europeans. The principal mines are in two valleys in which are the towns of Kathay and Kyatpyen.

Rubies and the minerals with which they are associated, such as spinel, are here found in a mother-rock of white, dolomitic, granular limestone or marble, of the upper Carboniferous age. These rocks have been altered by contact with molten igneous material which recrystallised the calcium carbonate as pure calcite, while the impurities became the ruby and its associated minerals. The precious stones are but occasionally found in the rock itself, but in an adjacent ground, which the miners call "byon," where the gem stones have weathered out; in the neighbouring river alluvium are found ruby particles, called ruby-sand. Prior to 1886 the rubies were mined by the Burmese with the primitive methods that had been in vogue for centuries, but when, in that year, Burma became part of the British Empire, the work was taken up first by an Anglo-Italian and then by an English company, which paid the Indian Government for this concession of mining rights the equivalent of about $125,000 annually.

Siam has long produced corundum rubies, but the gems are usually darker and inferior to the beautiful clear red stones from Burma. The principal mines are controlled by an English company. A few rubies have come from the gem-sands of Ceylon; a few have been found in Mysore and Madras, India; and inconsiderable products in Afghanistan and Australia. Rubies have been found in North Carolina and

Montana in the United States, but the products are not of commercial importance.

Corundum rubies formed of ruby material by artificial methods have attracted attention and are cutting some figure in the jewelry trade, but they are not and can never be the peers of natural rubies; man's ingenuity and science cannot compete with Nature in the gem business. Artificial rubies are described in another chapter.

Of the other stones than corundum called "ruby," the only important ones are the varieties of spinel, which chemically is closely allied to corundum so that the red varieties of spinel might be regarded as cousins-german to the real ruby. The "Cape ruby"—so called in the jewelry trade—is pyrope garnet from the diamond-bearing rock of South Africa, and is described in its proper place—the chapter on the garnet. Stones sometimes substituted for the ruby by dealers, or mistakenly called rubies, are red tourmaline, or rubellite, called "Siberian ruby"; rose topaz, called "Brazilian ruby"; and hyacinth or jacinth, which is zircon, and is described in the chapter on "Semi-Precious Stones Occasionally Used." Spinel has perhaps a wider range of colour than almost any other mineral, but it will be considered here chiefly with regard to the red varieties approximating the colour of the ruby. Spinel is practically a magnesium aluminate, consisting of alumina, 71.8%, and magnesia, 28.2%. The chief red shades are: deep red, Siam ruby and spinel ruby; rose red, balas ruby; yellow or orange red, rubicelle; violet red, almandine ruby. The native name in India for spinel is "pomegranate." A slight knowledge of mineralogy should suffice to distinguish the corundum ruby from its spinel distant relative, for the latter is less hard and of lower specific gravity, and different in crystallisation. Spinel is of about the hardness of topaz, or 8 in the Mohs scale, and its specific gravity is about 3.6. It crystallises in the isometric system and usually appears in the form of octahedrons. It is singly refracting, corundum doubly. Spinel is infusible before the blowpipe, but heating it will cause it to undergo several changes of colour, ultimately returning to its original hue, so that it might be termed the chameleon of gem minerals. Without any design to substitute spinel for corundum rubies, spinel has its own deserved value, and its beauty and intrinsic worth deserve for it an inclusion in the company of the high-class gems.

It is interesting to note that spinel ruby is not only the relative of the patrician corundum ruby, but the poor relation dwells together with its wealthy relative in nature. Both rubies are found associated in the gem gravels of Ceylon, Siam, Australia, and Brazil, as well as in the crystalline limestone of upper Burma. Spinel rubies are found in quantity in Balakschan, Afghanistan, near the River Oxus; the name "Balas ruby" is probably derived from Beloochistan, otherwise Balakschan.

CHAPTER VII

THE SAPPHIRE

SAPPHIRE, the stone of April, is the symbol of constancy, truth, and virtue. Like the ruby, it is corundum, and the name "sapphire" is generally applied to corundum of any colour excepting the red. More specifically, the name is applied to blue specimens, the desired tints being royal blue, velvet blue, and cornflower blue. A characteristic of this variety of corundum is, that occasionally its colour effect by artificial light differs from that manifested in natural light, being generally less brilliant. Dealers call the blue corundums "Oriental sapphires." It is one of the most ancient of stones and its names differ but slightly in the ancient languages, Chaldean, Hebrew, Greek, and Latin, from which the English word is derived. Stones of darker colour are frequently termed male and those of lighter shades female.

Higher specific gravity and a greater degree

of hardness, besides the difference in colour, distinguish the sapphire from the ruby; otherwise the sapphire's chemical and physical characteristics are generally included in the description of corundum in the foregoing chapter, covering the red corundum and other red stones termed rubies. While the form of the sapphire crystal corresponds with that of the ruby, there is a difference in the habit of crystallisation; the prism and rhombohedron of the ruby, are replaced in the sapphire by the hexagonal pyramid. The colouration of sapphires is frequently irregular; different portions of the same stone show different colours, and sometimes the body of what would be a colourless sapphire shows blue patches; but as the blue colour vanishes when the stone is heated, such a stone, undesirable as a gem, can be rendered valuable by heating it until it becomes a clear white sapphire. The colours of sapphire range from the white, colourless, or, so-called, "Leuco-sapphire"; through the yellow, called "Oriental topaz"; and through various tints to the royal blue of the typical gem sapphire. Sometimes sapphires show different colours at their terminations, as greenish-blue at one end and blue at the other, or red and blue at the ends; examples have been seen that were blue at the ends and yellow in the middle. One famous tri-coloured sapphire is cut into a figure of the Chinese sage, Confucius; the head is colourless, the body pale blue, and the legs yellow. Sapphires exhibit as many shades of blue as can be named. The very darkest, almost black, is termed "inky"; pale "feminine" stones are termed "water-sapphires"; dark, yet very blue stones, are called "indigo-sapphires," "lynx-sapphire," or "cat-sapphire." The tone and transparency of the stone are most important factors, and, provided they are present, the very dark shades are not disadvantages, although the "corn-flower" is the choicest. Besides the "corn-flower" colour, tones and tints are indicated by such adjectives as "Berlin," "smalt," "grey-ish," and "greenish." The dichroism of the sapphire is nearly always apparent if the stone is viewed from an angle that reveals it, the blue appearing tinged with green or with violet. The dichroism of the sapphire is, like that of the ruby, taken into account in producing the best effects in the cutting. In artificial light some specimens remain unchanged, while others be-come darker, or, perhaps, change to a reddish, purple, or violet colour. Asterias or star-

CHAPTER VIII

THE AMETHYST

THE amethyst is a species of quartz that is now of more artistic than intrinsic value. The native beauty of the purple stone is indisputable. In folklore it has a prominent place as the natal stone of those born in the month of February, who, astrologically dwell in the sign of Aries in the Zodiac, and are dominated by the planet Mars. It is distinctively the precious stone of the Bishop, and also, rather incongruously, of Bacchus; and yet, despite its appropriation by these personages, respectively ecclesiastical and mythological, it is also used as an amulet believed to protect the wearer from the curse of excessive indulgence in stimulating beverages. The amethyst is the symbol of pure love; it is also the "soldier's stone"; it is the stone appropriate for mourning, and thus, in many ways it is invested with a strong sentimental interest. The signet ring of Cleopatra was an amethyst, engraved with the figure of

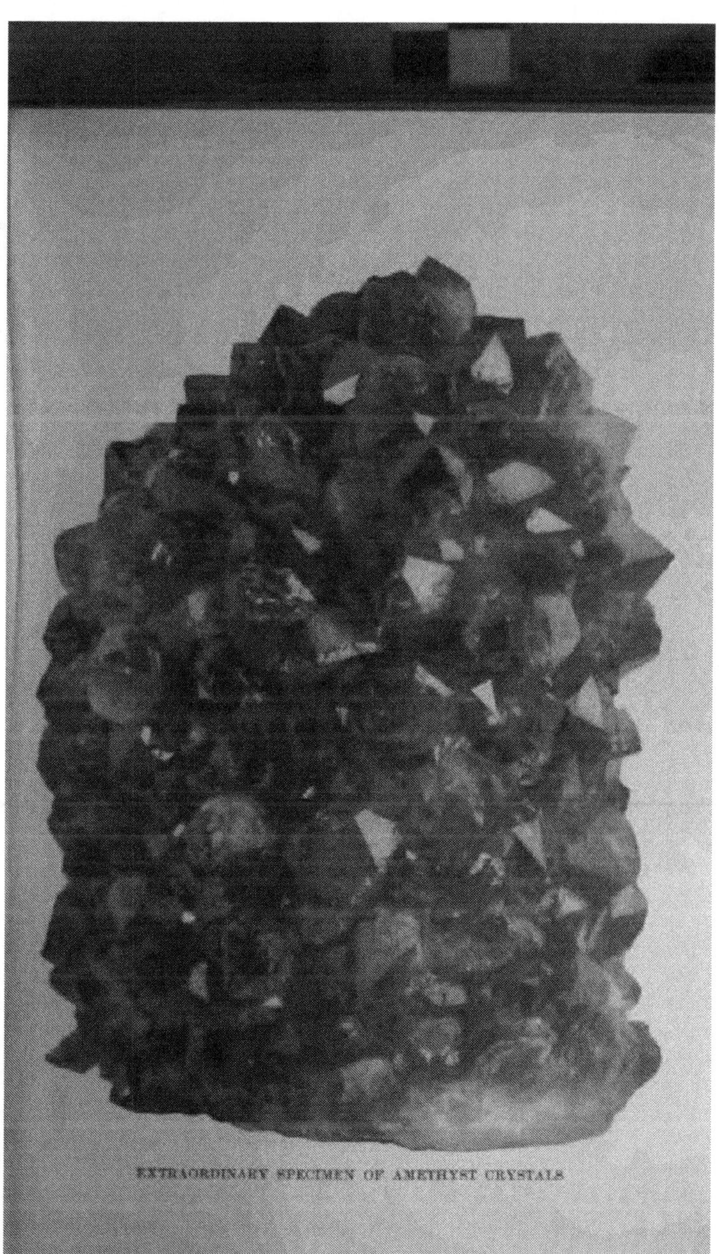

EXTRAORDINARY SPECIMEN OF AMETHYST CRYSTALS

Mithras, a Persian deity, symbol of the Divine Idea, Source of Light and Life. From the ring of Edward the Confessor was taken the amethyst that adorns the British crown, and this particular stone is, by tradition, imbued with the qualities of a prophylactic against contagious diseases.

There is an ancient myth that a beautiful nymph was beloved and beset by Bacchus, who in her effort to escape the imperious wooing of her ardent lover, was aided by her patron goddess and metamorphosed into an amethyst. Bacchus, baffled, in memory of his vanished love, bestowed on the stone the colour of the purple wine he best loved, and registered a vow that forevermore whoever would wear the amethyst should be preserved from intoxication, no matter how extensive his libations. In medieval times the amethyst was a favourite amulet as a preserver of the wearer in battle, and many a pious crusader who nightly told his beads, relied also upon the purple stone that hung as a protective charm beside his rosary. The amethyst was believed to be a good influence if worn by persons making petitions to princes, and also to be a puissant preventive of hailstorms and locusts. The association of the amethyst with sacerdotal things is old and long, for it is the pious or episcopal gem, and regarded as imparting especial dignity and beauty to the property of the Roman Church. The amethyst is sacred to St. Valentine, who is said to have always worn one.

The word amethyst owes its root to the Greek word *amethustos*, meaning not drunken, and also construed to mean a remedy for drunkenness. Pliny, with customary quaintness, thought it prevented intoxication because it did not reach, although it approximated, the colour of wine.

Amethyst, a variety of quartz, plainly crystalline, is called by Dana, amethystine quartz. Its colour, which is diffused throughout the crystals or affects only their summits, is a clear purple or bluish-violet, and it is therefore sometimes called violet-quartz. The amethyst is of all degrees of colour from the slightest tint to so dark as to be almost opaque. Not always uniform, the colour is sometimes in spots and in some crystals shades gradually from light to dark. The dark reddish-purple colour is most highly prized; it has the advantage, too, of holding its value under all circumstances, for in an artificial light, especially if containing yellow rays, the pale stones lose their violet colour and become a dull grey. Some deeply coloured amethysts from Maine change to a wine colour in artificial light, thus becoming even more beautiful.

The amethyst's best claims to perpetual popular appreciation are its beauty of colour and its adaptability as an ornament to harmonise with a costume colour scheme. In the development of woman's discrimination in dress, she desires a jewel for every gown and ornaments for afternoon as well as for night, and for special occasions. For fabrics of pearl-grey, amethysts mounted in dull silver should be in high favour.

A good amethyst should be of a deep purple colour, perfectly transparent and throughout uniform in hue. Amethysts are distinctly dichroic; they rank No. 7 in the Mohs scale of hardness; specific gravity is 2.6 to 2.7. The crystallisation of this quartz is in six-sided prisms terminating in pyramids. Lustre vitreous; cleavage none or distinct; fracture conchoidal, glassy. It is doubly refractive, the twin colours being reddish and bluish purple. Amethysts are usually cut step, while the finer specimens are cut brilliant.

The chief sources of supply for amethysts are Brazil and the Ural Mountains, Siberia. The Siberian amethysts, accompanied by beryl and topaz, occur in cavities in granite; often they are found lying loose and sometimes very near the surface. Cavities in a black eruptive rock (melaphyre) are the hiding places of some Brazilian amethysts, while others are found as pebbles in the river gravels with chrysoberyl and topaz as companion minerals. Gem amethysts are also found in gravel bearing other gems in Ceylon.

In North America, a few of the finest specimens of amethyst on record have been found in Oxford County, Maine. Other localities are Delaware and Chester Counties, Pennsylvania, and Haywood County, North Carolina. Crystallised amethyst in commercial quantities has been found at Thunder Bay on the north shore of Lake Superior. The crystals are highly coloured but not uniform or clear and few good gems have been obtained there.

Amethyst was formerly much more highly prized than now because of its scarcity. Besides the increased supply it has been imitated so convincingly as to impose upon all excepting gem experts. A celebrated amethyst necklace

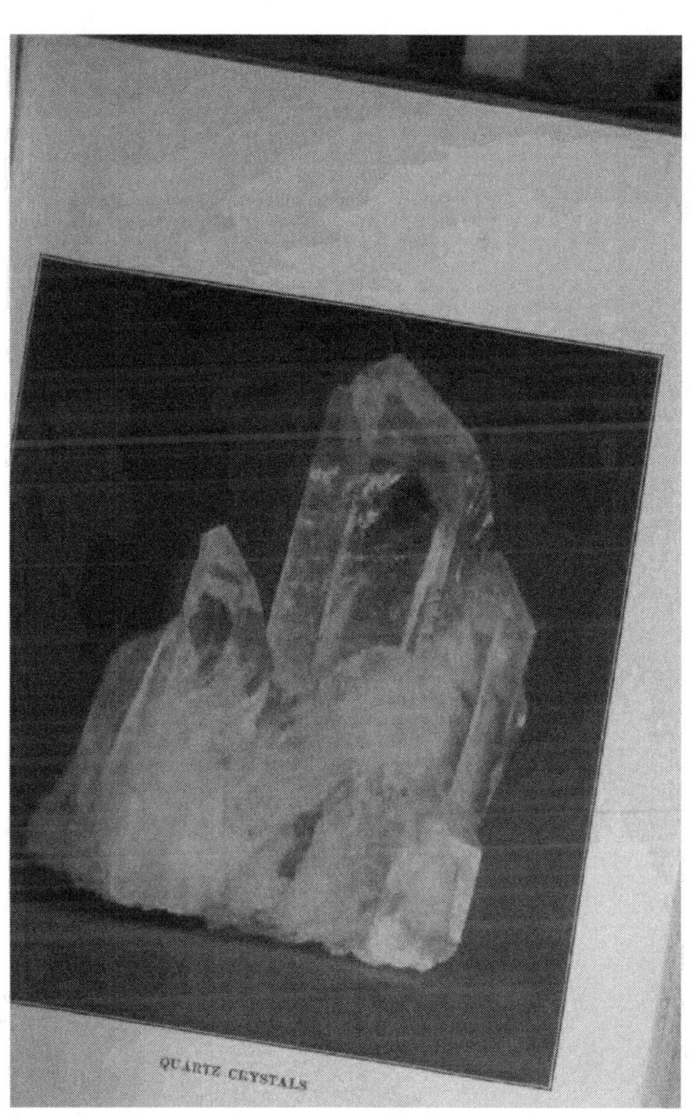

QUARTZ CRYSTALS

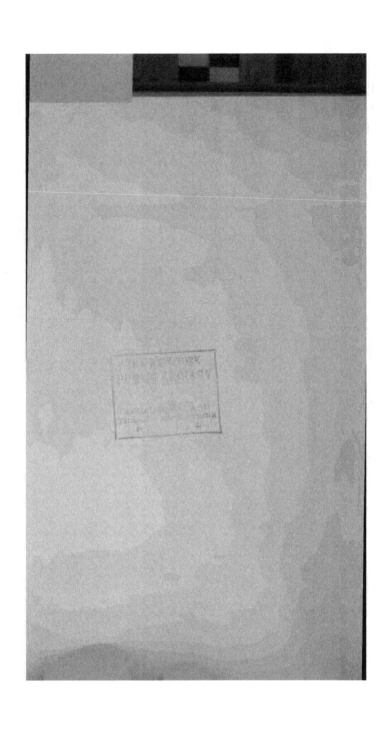

owned by Queen Charlotte of England, valued at $10,000, might not now be worth intrinsically $500. The exclusive charming violet colour of the amethyst will probably always insure a demand for the best qualities of this stone, and with a development of art in the treatment and uses of precious stones and jewelry, the demand is likely to grow.

Of all specimens of amethyst that appear in the market to-day, the Siberian stones so far outclass all competitors in richness and depth of their dark violet hue, that these beautiful gems mixed with others would be instantly selected by the merest novice; so manifestly superior is their quality that, comparatively speaking, they alone are gems, and the only reason that their cost is not much greater than it is, is because Nature has been generous in the quantity that she has permitted man to extract from her mineral treasure house.

CHAPTER IX

CORAL

CORAL has been used for personal ornamentation, and as an article of commerce, from the earliest period recorded in writing. Popular to-day, as it has almost always been—especially in the form of polished fragments, pierced and strung like beads, and less extensively in beads, spherical or oval—the most desired, high grade of light rose-pink coral is becoming scarce, and those who gather it from the ocean's floor are anxiously seeking new sources of supply. At the present time coral is increasing in favour and the demand for it is steadily growing.

Coral—like the sea gem, the pearl,—is essentially carbonate of lime. Its structure is erected by a family of zoöphytes, gelatinous marine animals (not insects as is too often written) called polyps. The coral is secreted by a peculiar layer of the skin; it is the calcareous skeleton of the lowly organised animal, and gradually develops like the bones of vertebrates, and is not built up as bees build a honeycomb as is popularly believed. The pits or depressions on a branch of coral represent the places where the coral colonists once grew. Coral is a common submarine feature in low latitudes all around the globe, but the gem or precious coral, *Corallium rubrum,* formerly called *Corallium nobile,* comes almost exclusively from the Mediterranean Sea off the African, Corsican, and Sicilian coasts. A wild-rose pink is the particular shade most highly favoured. The *Corallium rubrum,* the only species utilised and valued to any extent for jewelry, belongs to the family Gorgonidæ of the group Alcyonaria.

The skeleton of a colony of *Corallium rubrum* is found to be cemented firmly by a disc-shaped foot to any dense natural or foreign object on the sea bottom, as a stone, cannon-ball, bottle, or, as is recorded in one case of fact, a human skull. The branches seldom exceed a foot in length and an inch in diameter. A curious characteristic of coral is, that it grows always perpendicular, or approximately, at a right angle to the surface to which it is attached—downward, if its foothold is on the under face of a rock.

The colonies are usually from sixty to one hundred feet beneath the sea's surface.

Some expert authorities have fathered the assertion that about thirty years is required for coral stock to develop into full size; yet the Sicilian coral bank is divided into ten sections, one of which is finished every year, and at the end of the decade the first bank yields full-sized stock.

Pietro Moncadi of Palermo, said to be the largest dealer in Italian coral, during a recent visit to New York, reported that the demand for high grade red coral leads the supply. Many beds off the Italian coasts are exhausted and there is much prospecting off Malta, Malabar, and East African coasts, at great expense and, so far, with very small reward. Signor Moncadi made the statement that the United States buys the finest red coral, and the producers who possess the highest grade have to seek no other market.

The home of the coral industry is Italy, where there are about sixty work shops, with about six thousand employees. Torre del Greco is the centre of both the coral-fishing and the coral-working industries. The coral-workers pierce and string pieces of coral of all shapes and sizes. The beads are spherical or egg-shaped—the latter are called "olives." The handicraft of the Italian coral-workers includes carving of a high artistic order—the forms representing many natural objects—and the cutting of beautiful cameos. The coral-gatherers employ fine distinction in denominating coral tints. Pure white is *bianco,* fresh pale flesh-red is *pelle de angelo;* pale rose, *rosa pallido;* bright rose, *rosa vivo;* these choicest tints are followed by "second colour," *secondo coloro;* red, *rosso;* dark red, *rosso scuro;* and, darkest of all reds *carbonetto* or *ariscuro.*

The specific gravity of precious coral is 2.6 to 2.7; hardness in Mohs's scale about 3–4. Coral is soft enough to be easily worked with a file, edged tools, and on a lathe; it is too soft to take a high polish, but despite that dissimilarity from the precious stones of whose company it is a popular member, its fine colour sustains its claim to beauty, and it highly deserves inclusion in a book of gems.

But little coral, comparatively, is mounted in Italy, the setting being done in the fashion in demand in the country where it appears in the jewelry trade.

In the Orient coral is always in demand, with

India in the lead followed by China and then Persia. The Chinese mandarins sometimes pay incredible sums for exceptionally fine coral buttons for their caps.

Pieces of coral are used for rich and costly handles of parasols and umbrellas; the coral handle of an umbrella belonging to the Queen of Italy being valued at nearly two thousand dollars. A coral necklace exhibited in 1880 at the International Fisheries Exhibition held at Berlin, was valued at nearly twenty-nine thousand dollars. In Italy the superstition that the wearing of coral is a protection against the evil eye, accounts for its appearance as the commonest personal ornament among the masses; similarly, it is in evidence among the lower class of Italians in the United States. Coral is easily imitated, however, and most of the defences thus relied upon by superstitous wearers are spurious, but equal to the genuine in efficacy. Red gypsum is a common sophistication for precious coral, and simple tests are: scratching it with the finger nail and the application of acid, under which it does not, like genuine coral, effervesce. Celluloid is now sometimes used as a substitute for coral.

The existence of coral within the United

States, on the shores of Little Traverse Bay, at Petoskey, Michigan, should not escape mention in an American book. The coral found here is fossil, and many specimens possess rare structural beauty; they are compact and susceptible to a high polish. The fragments found are water-worn, and the weight of some masses secured attained to three pounds. The colour is grey, of various shades. Local lapidaries cut and polish these handsome fossil relics of a prehistoric submarine period, and shape them into seals, charms, cuff buttons, and paper weights. In the mineralogical section of the reports on the Eleventh Census, 1900, Mr. George Frederick Kunz records that from $4000 to $5000 worth annually were sold.

CHAPTER X

GARNET

GARNET is a noun that is applied to a variety of gem minerals red or brownish-red. Almandite, a stone of rich cherry, claret, or blood-red colour is the precious garnet. A variety of garnet recently established that is in high favour is rhodolite. The chemical bases of both of these leading varieties are the same, a silicate of iron and aluminium. Precious garnet has a hardness of about 7.5, with a specific gravity seldom less than 4. and occasionally as high as 4.3. Closely following almandite, or as jewellers call it, "almandine," in the favour of gem fanciers, is Bohemian garnet or pyrope, meaning "fire-like"; this has a range of colour from a deep blood red to almost black. Pyrope is slightly harder than almandite, and its specific gravity lies between 3.7 and 3.8. The fracture is brittle; refraction, single; lustre,

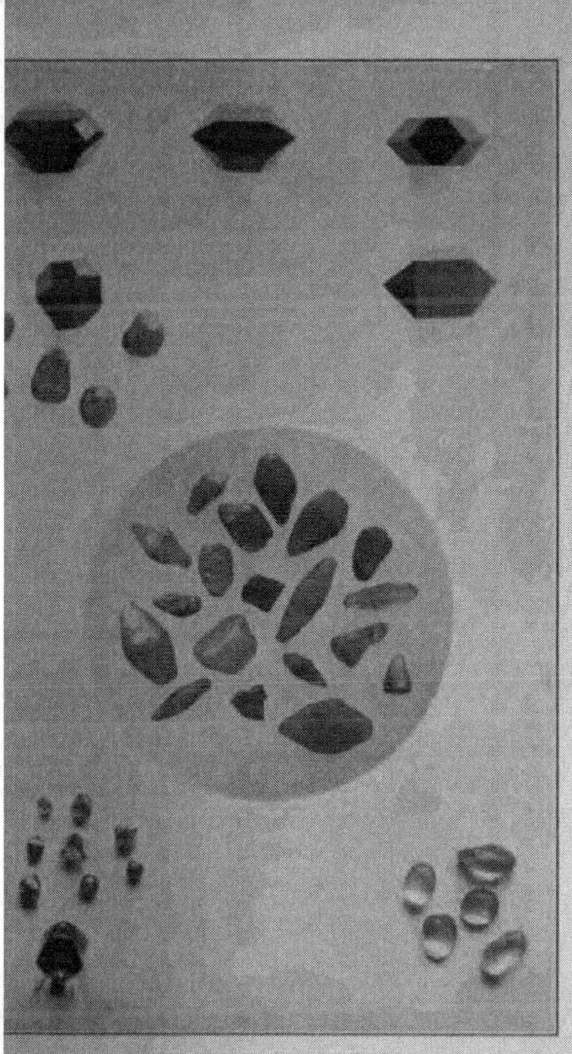

GARNET CRYSTALS AND PEBBLES OF PYROPE
SAPPHIRES
IAMOND CRYSTALS FROM KIMBERLEY MINES, SOUTH AFRICA
Specimens in U. S. Nat. Museum

THE NEW YORK
PUBLIC LIBRARY
ASTOR, LENOX
TILDEN FOUNDATION

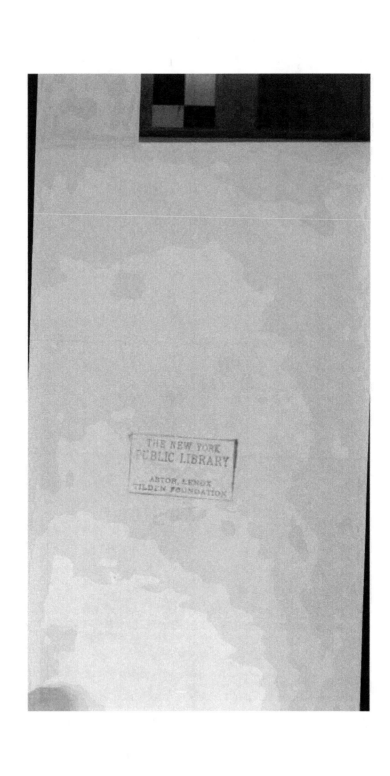

vitreous; it is transparent to opaque. Most varieties of garnet fuse to brown or black glass.

In Dana's *Mineralogy*, Garnet is *Carbunculus dodecahedrus:* order Hyalina. In crystallography the primary form of garnet is the rhombic dodecahedron. The cleavage is indistinct parallel with the faces of the dodecahedron. Besides the primary twelve-sided form, with rhombic faces, the secondary forms of garnet crystals include trapezohedrons—twenty-four-sided forms—with faces shaped like trapeziums; then there are combinations of these forms, one of which has thirty-six faces. The tendency of garnet is to crystallise and it is usually found in crystals; these range from tiny ones the size of a grain of sand up to those of several pounds in weight.

The name garnet, according to one version, is derived from the Latin *granatus*, meaning like a grain, because of the resemblance of its crystals in size and colour to the seeds of the pomegranate.

A carbuncle, in the popular conception, is a specific precious stone, but it does not exist in scientific mineralogy, and in the verbiage of dealers now, its meaning is merely any worthy red translucent stone cut *en cabochon*. Some writers, who seem otherwise generally well informed, have fallen into this common error of recognising the word carbuncle as the name of a specific gem. Probably almost any fiery-red translucent ornamental stone in the days of ancient Rome was called *carbunculus*, derived from *carbo*, coal, and the name was bestowed because of the internal fire-like colour and reflection which is a common characteristic of the various stones now generally termed garnets. The garnet is among the stones earliest mentioned in the surviving literature of all ancient languages.

Almandite derives its names from Alabanda, a city in the ancient district of Caria, Asia Minor; whence garnets were introduced to ancient Rome. The most highly valued specimens of almandite, for a long period, came from localities not known to the western world, but they were supposed to be mined near the city of "Sirian" in old Pegu province, Lower Burma, and were called "Sirian garnets." So careful an investigator and high an authority as Dr. Max Bauer, in his monumental work on precious stones, states that Syriam, the ancient capital of Pegu, is now but a small village in the British province of Lower Burma near the great trade centre of Rangoon. A résumé of the facts evolved by Dr. Bauer shows that no precious almandite occurs in any part of Burma, while in Upper Burma the only red stones found are ruby, spinel, and red tourmaline. Long ago, therefore, Syriam was merely a distributive point for garnets brought to its market from a distance, possibly from the Shan states to the eastward. The "Sirian" garnet is now merely a type; it tends toward a violet colour.

In northern India almandite is mined on an extensive scale in several localities. The stone is found in the Alps, Australia, and Brazil; a variety too opaque to be very valuable, occurs plentifully on the Stickeen River in Alaska. Metamorphic rocks, such as gneisses or mica schists, granite, and gem gravels are the usual environments of almandite.

Rhodolite is an intermediate between almandite and pyrope, more closely related to the latter, but differing in colour from both. It is found as water-worn pebbles in the gravels of Cowee Creek and Mason's Branch, Macon County, North Carolina; sometimes it occurs along with ruby in a decomposed, basic igneous rock, known as "saprolite"; a curious occurrence is in the form of small crystals enclosed in crystals of ruby. The colour resembles that of the rhododendron, from which this but recently recognised precious stone was christened rhodolite. Although mineralogically different from almandite, and more like pyrope, rhodolite is known in the trade as "almandine," and, in the United States at least, is bought and sold under that title; the difference in composition and colour is too slight for merchant jewellers to recognise, and the name "rhodolite" is scarcely known to the trade or the general public. In fact, in the jewelry trade, any garnet with a tendency toward a violet colour is classed as an "almandine." Under the name "almandine," there has been an increased demand for this variety of garnet for medium-priced jewelry for about five years previous to this writing.

Scarcely second to almandite, is the dark blood-red pyrope, found in company with the diamond in South Africa, and, in the trade, called "Cape Ruby." This fine South African gem stone, companion of the diamond and native to the world's greatest diamond fields, is a magnesium-aluminium garnet, containing manganese oxide and ferrous oxide; its specific gravity is 3.86, approximating that of the Bohemian pyrope, which it resembles in both chemical composition and colour, thus clearly classing it as pyrope, and not almandite, as was done for some time after its discovery. In the trade at present this variety of garnet commands a higher price than any other.

Varieties of the lime-aluminium garnet occasionally appear in gem-stone commerce. Lime-aluminium garnet has a hardness of 2.7, and a specific gravity of 3.55 to 3.66. Its colours are white, pale green, amber, honey, wine, brownish-yellow, cinnamon, brown, and pale rose-red. The varieties include essonite and cinnamon stone, the latter often improperly called, by merchants, "hyacinth." The gem cinnamon stones come chiefly from Ceylon; they are of a cinnamon brown, or range from that to a deep gold colour tinged with brown. Grossularite includes the pale green, yellow to nearly white, pale pink, reddish or orange, and brown kinds. Romansovite is brown. Wilnite is yellowish-green to greenish-white. Topazolite is topaz, to citrine, yellow. Succinite is amber coloured. There are two kinds of calcium-iron or green garnets: The demantoid, from the Ural Mountains, Siberia, has a hardness of 6. to 6.5; specific gravity, 3.83 to 3.85. Demantoids have a rich green colour and when clear and flawless are beautiful lustrous gems; the choicest are called "olivines." The other green

8

variety, Uvarovite, is found chiefly in Russia. Montana ruby is a trade term for the fine garnets found in Montana and Arizona. The finest American garnets are found in the territory of the Navajo nation in north-western New Mexico and north-eastern Arizona, where they are collected from ant-hills and scorpions' nests by Navajo Indians and sometimes by United States soldiers from adjacent forts. According to the most eminent authority on American gem stones, Dr. George Frederick Kunz, these red stones, known locally as Arizona and New Mexico rubies, are unsurpassed, equalling in value those from the Cape of Good Hope. Fine gems weighing two and three carats, after cutting, are not rare. By artificial light the American stones are superior to "Cape rubies." These American garnets have evidently recently weathered out of a peridotic rock.

Another type of garnet is known as spessartite, a variety of essonite, in which part of the alumina is replaced by manganous oxide. The finest specimens of this variety known were discovered at Amelia Court House, Virginia, which locality has yielded gems weighing from one to one hundred carats.

CHAPTER XI

THE OPAL

THE precious opal is one of the most individual of gems; of all the opaque minerals, it reveals the most beautiful play of colours, in folklore it is the birth-stone of October and the symbol of hope, and yet, for years, the fame of this fire-flashing stone was blackened by a cloud of superstition which condemned it as unlucky; a superstition the origin of which is obscure. For a time, however, it largely regained its lost popularity, having found its most illustrious patron in Her Majesty, the late Queen Victoria. Another remarkable fact about the opal is that it is not found in the Orient—the very land of gems.

Opal, in mineralogy, is *Hyalus opalinus*, of the order Hyalina; it is of granular structure; small reniform and stalactitic shapes and large tuberose-like concretions; hardness 5.5 to 6.5; specific gravity 2 to 2.21; lustre vitreous, sometimes inclining to resinous or pearly; streak, white; colour, white, yellow, red, brown, green, or gray. The colour is usually pale, due to foreign elements. Some opals exhibit a rich play of colours, while others present different colours by refracted and reflected light. The cause of the colour-play is the physical condition resulting from a multitude of fissures having striated sides which diffract and decompose the light. The chemical composition of the opal is ninety per cent. silica and ten per cent. water.

Besides precious opal, there is the harlequin opal which presents a variegated play of colours on a reddish ground, and resembles the fire opal which shows hyacinth red to honey-yellow colours, with fire-like reflections. Girasol is bluish-white and translucent, and, under a strong light, presents reddish reflections. Lechosos opal is a variety remarkable for flashes of green. Hydrophane, a light coloured opaque kind, becomes transparent when immersed in water. Cacholong is an opaque porcelain, bluish, yellowish, or reddish white. Opal agate has an agate-like structure. Jasp opal contains iron, and is to opal as jasper is to quartz. Wood opal is wood silicified by opal. Hyalite (Müller's glass) is colourless and clear, or translucent and a bluish white. Moss opal contains manganese oxide, and is to opal as moss agate is to quartz. A freakish variety of opal is tabasheer, a silica deposited within the joints of bamboo; it is absorbent, and, like hydrophane, becomes transparent when immersed in water.

As a mineral, opal is quite common, so that an amateur's collection of minerals can include specimens to represent opal—some of them very beautiful, too—at small cost, or for the effort of prospecting, in many localities. The varieties of opal are many, and the frequent inclusion of foreign matter invests it with a wonderful variety of colours. The silica deposited by nearly all natural hot waters is opalescent. The Yellowstone Park geysers shoot up around cones of opal raised by the constant accretions of silica deposited by the passing hot waters, which fall into opal basins created in the same way. This variety of opal is termed geyserite. There is a wide gulf in values between precious or noble opal—the gem stone quality—and opal in general.

Opal is generally found filling seams, cavities, and fissures in igneous rocks, also embedded in limestone and argillaceous beds.

Opals of a quality fit for use as ornamental

stones are found in many lands. Mines in Czernowitza, in northern Hungary, long produced the most highly valued gem opals obtainable. These opals are often known as "Oriental opals," because they first appeared in Holland through Greek and Turkish traders. Despite the trade practice of applying the term "Oriental" to this type of opals, none is found in the Orient. The Hungarian opals were undoubtedly those first known to the Romans. The claim is made that Hungarian opals are less likely to deteriorate than any other variety. Gem opals are also found in Australia, Mexico, and Honduras. Although opals are produced to a commercial extent in several Mexican states, they are most systematically mined in Queretaro, where the opal occurs in long veins in a porphyritic trachyte. This opal mining has created a somewhat primitive cutting and polishing industry in the city of Queretaro. The exporting of Honduras opals—all uncut—is not extensive. In the United States the occurrence of gem opal has been observed in the John Davies River, Oregon, and near Whelan, between the Cœur d'Alene and Nez Percés Indian reservations, almost on the Idaho line, State of Washington. The most prolific source of opals in recent years has been the Australia mines, the most prominent being White Cliffs, New South Wales. Extensive mining operations are carried on there, the matrix of the opal being a cretaceous sandstone, which has been permeated by hot volcanic waters. The output of this region has already been represented by millions of dollars. Opals have been obtained in commercial quantities at localities on the Barcoo River and Bulla Creek, Queensland, and are occasionally found in West Australia.

The admiration of the ancients for the opal is expressed by Onomacritus, writing five hundred years B.C., who remarks: "The delicate colour and tenderness of the opal remind me of a loving and beautiful child." Pliny, whose voluminous books covered so wide a range, and who evidently believed himself qualified to write about anything, wrote of the opal: "It is made up of the glories of the most precious gems, and to describe it is a matter of inexpressible difficulty." The ancients esteemed the opal highly, and attributed to it an influence for every possible good; this belief outlasted the Middle Ages, and in the early part of the seventeenth century the opal is recorded as being as highly valued as ever. Then arose a superstition that the fiery stone was unlucky, and this became prevalent everywhere. The cause of this has been attributed to Walter Scott's novel *Anne of Geierstein*. A genuine reason why opal may have come to be regarded as unlucky by its possessors is its mutability. The changes which may occur in the opal are not only numerous but freakish and uncanny. Brilliant opals have lost their fires and lustre forever, while others have lost and recovered them. In other cases dull specimens have suddenly developed brilliancy. Mediocre specimens will sometimes, when moistened with oil or water, exhibit a fine colour play, which will vanish when the stones dry, and this peculiarity has been utilised for profit by dishonest dealers. A stone thus acquired would be unlucky for the purchaser.

Credit for the reinstatement of the opal in public favour is believed by the author to be due in great part to the late Queen Victoria, who, in many ways, demonstrated her royal favour for the stone of many fires and colours, and there is no doubt that the Queen's motive was to benefit her colonial subjects in Australia, where opals had been discovered.

Queen Victoria gave to each of her daughters,

at their marriage, opals, and this and other acts which signified her admiration for the stone and her disdain for the superstition through which its reputation had fallen into evil days soon raised opals high in the realm of precious stones; the result being that Australian exports of opal were handsomely increased by the demand for, and shipment of, the stone thus royally reinstated to its ancient high estate of popular favour.

It is but a just appreciation of the average high intelligence of the gem-purchasing American public, to state that opals have always been appreciated in the United States for their merits, and that here the dread tabu of "unlucky" has had the least effect. And it may be said that it is on their merits they are judged, for the demand has latterly distinctly decreased for the inferior grades of opals that formerly sold readily, while choice gems are sought for, and American purchasers prove themselves well posted and very discriminating.

CHAPTER XII

THE TOPAZ

YELLOW is the colour generally associated with the topaz, yet topaz is sometimes colourless, or may present almost any colour, and beautiful specimens of other colours are often supposed to be some other mineral, so thoroughly identified is this stone with the colour yellow. The sometime popularity of topaz has of late years declined, and a probable reason is the common substitution of other stones for it. Topaz takes its name from *Topazios*, meaning "to seek"; because the earliest known locality from whence it came was an island in the Red Sea which was often surrounded by fog, and therefore difficult for the local mariners to find.

The name of topaz in mineralogical science is *Topaz rhombicus*, and, like the opal, it belongs to the order Hyalina. The primary form of topaz in crystallography is a right rhombic prism. Its cleavage is parallel to its basal

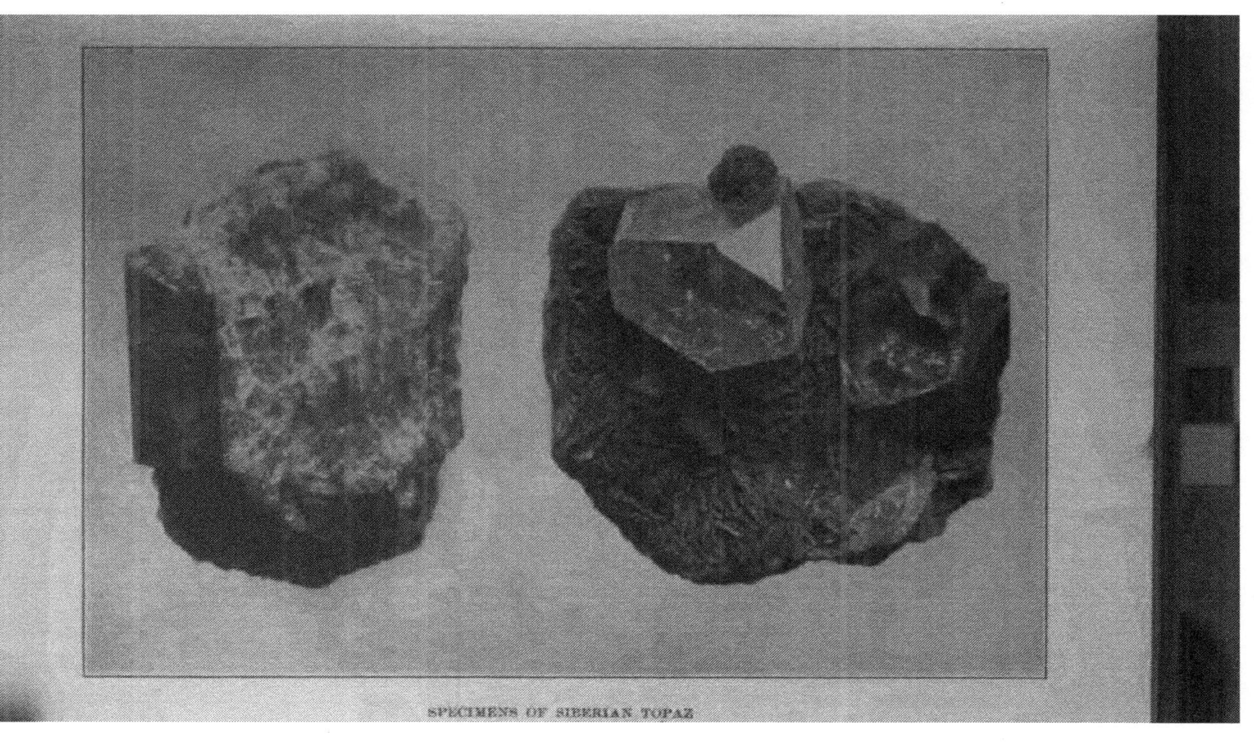

SPECIMENS OF SIBERIAN TOPAZ

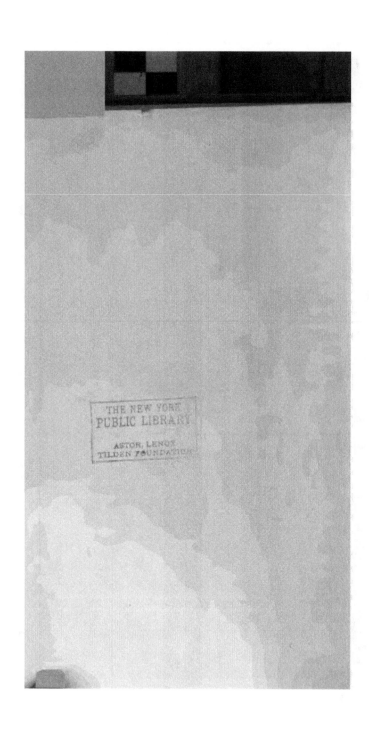

plane, almost perfect, and it cleaves so easily that a cut topaz, if dropped, might be easily cracked or broken. The crystallisation of topaz is imperfect; structure, columnar; lustre, vitreous; streak, white. Topaz is either transparent or translucent; the colours of topaz including wine, amber, honey, and straw-yellow, pale blue to pale green of many shades, greyish, reddish, and white. Rolled pebbles of limpid colourless topaz are called by Brazilians "*pingas d'agoa*," and by the French, "*gouttes d'eau*," both meaning drops of water. The coloured varieties show marked pleochroism. The fracture of this mineral is conchoidal and uneven.

True topaz is a silicate of alumina, containing hydroxl and fluorine; hardness, 8; specific gravity, 3.4 to 3.6. Being three and one half times as heavy as water, topaz can be readily distinguished from other stones resembling it by those accustomed to handling them. Topaz cannot be fused on charcoal before the blowpipe, but it is partially decomposed by sulphuric acid. Its hardness enables it to take a high polish, and the colourless variety has been cut in brilliant or rose form so as to resemble the diamond, for which it might readily pass in daylight. However, it is but weakly doubly refractive and dispersive, and its comparative softness makes its distinction from the diamond a simple matter. Although infusible, when sufficiently heated, the faces of crystallisation of topaz become covered with small blisters which crack as soon as formed; and with borax it slowly forms a clear glass. Some varieties assume a wine yellow or pink tinge when heated. The rose-pink topaz sometimes appearing mounted in jewelry, is not natural; the delicate tint of this gem with an artificial complexion results from a simple process called "pinking," applied to yellow or brown kinds. A topaz selected to be "pinked" is packed in magnesia, asbestos, or lime, and carefully and gradually heated to a low red heat; the stone then being slowly cooled. If the temperature attained has not been sufficiently high, the desired rose-petal tint is not obtained and a salmon tint appears; if the temperature rises too high, or is too long continued, the colour completely disappears. Pulverised topaz changes to green the blue solution of violets. Topaz generally becomes electric by heat, and if both terminations of the subject specimen are perfect, polarity will be developed; transparent varieties are susceptible to electrical excitation by friction.

Several minerals are commonly called topaz; yellow sapphire is called "Oriental topaz"; and varieties of quartz are called "Saxon," "Scotch," "Spanish," "Smoky," and "False" topaz. The hardness, weight, and power of developing frictional electricity, possessed by the true topaz, enable investigators to distinguish real topaz from these nominal varieties.

Topaz commonly occurs in gneiss or granite, associated with tourmaline, mica, or beryl, and occasionally with apatite, fluor-spar, and tin. The purest variety of topaz, perfectly colourless and pellucid, is not uncommon; as crystals it is found in Miask, in the Ural Mountains, Siberia, and, abundantly, as water-worn pebbles, in the river and creek beds of Diamantina and Minas Novas in the state of Minas Geraes, Brazil. Mineralogists regard the "Braganza," a gem claimed to be a diamond, included in the crown jewels of Portugal, and weighing 1680 carats, as one of these pebbles; probably one of the finest ever found. A sobriquet for these clear colourless topazes is "slave's diamonds." Blue topaz from Brazil is sometimes termed "Brazilian sapphire." A fine saffron-yellow variety, called "Indian topaz," occurs infrequently in Ceylon, and rarely, in Brazil; the golden yellow tinted variety from Brazil is the kind distinguished in the jewelry trade as "Brazilian topaz." Schneckenstein, near Gottesberg, in the vicinity of Auerbach, Voigtland, Kingdom of Saxony, is said, by Dr. Max Bauer, to be the most important European locality producing topaz; it is there imbedded in a steep wall of rock, and occurs in small fragments of schists rich in tourmaline, cemented firmly into a hard mass by quartz and topaz. Brazil is the main source of topaz, and a review of the localities, association, and varieties of its established occurrence there would require an extensive space.

In North America topaz is found to an extent of small commercial importance in Mexico. In the United States it occurs more abundantly, although gem-quality is rare. Colorado has yielded the best specimens from localities in Chaffee County and El Paso County, on Cheyenne Mountain and elsewhere in the region of Pike's Peak. Small but brilliant crystals have been found at Thomas Mountain, Sevier County, Utah. At Bald Mountain, North Chatham, New Hampshire, topaz occurs, with phenacite, in crystals.

CHAPTER XIII

TURQUOISE

TURQUOISE is a popular gem mineral to-day, as it was anciently with the Persians and the Aztecs, whose name for it was chalchihuitl. Turquoise is a French word, meaning a Turkish stone, also the feminine of Turkish. Turquoise is an amorphous stone occurring in kidney-shaped nodules and incrustations; its colour is various shades of azure or robin's egg blue. Of Persian origin, it is supposed to be the stone anciently referred to, in Pliny's natural history, as callais, callaina, and callaica. In his catalogue of gems in the United States National Museum, Wirt Tassin applies to turquoise the names callainite and turkis; Cattelle says it is known to scientists as "callaite"; Oliver Cummings Farrington in his *Gems and Gem Minerals* describes callainite as a distinct mineral.

The hardness of turquoise is 6; specific gravity, 2.6 to 2.8; there is no cleavage; it is brittle and breaks unevenly. The lustre of turquoise is waxy and the colour is sky-blue, bluish-green, apple-green, and greenish-gray. The colour is liable to change, however, the blue becoming a pale green. Artificial means are resorted to for "improving" stones of a poor colour, but a washing in strong ammonia water will expose the fraud. This solution will not affect the colour of the true turquoise, but as soap and water does, possessors of rings set with turquoise should never wash their hands without removing their rings.

The chemical composition of the turquoise is a hydrous phosphate of aluminium and copper, and the principal components in a hundred parts are: phosphoric acid, 30.9; alumina, 44.50; oxide of copper, 3.75; water, 19.

The exposure of turquoise to a sufficiently high degree of heat will extract the water and cause it to crackle.

The turquoise most highly prized comes from Persia, and the most celebrated are those from an old mine, the Abdurrezzagi in a district of the Nishapur province in the north-eastern part of the country. Less valued specimens come from Asia Minor, Turkestan, and the Kirghiz Steppes. The Egyptians mined turquoise in the Wady Maghara, in the desert of Sinai. Specimens from Arabia in modern times proved of little value, fading quickly when exposed to the light. The mineral has also been found in Victoria and New South Wales. The United States is constantly growing in importance as a source for supply for the world market for turquoise. A trachytic rock in the Los Cerillos Mountains near Santa Fé, aboriginally worked by the natives, is a well-known mine, and some beautiful specimens have recently been found there. Other localities are Turquoise Mountain, Cochise County, and Mineral Park, Mojave County, Arizona; Columbus, Nevada; Holy Cross Mountain, Colorado; and Fresno and San Bernardino counties, California. Record specimens come from the mines of the Azure Mining Company, Burro Mountains, New Mexico.

Because of the opacity of turquoise, it is seldom cut with facets, but in a round or oval form, with convex surface; as the pieces suitable for cutting seldom reach a large size big turquoise gems are almost unknown. Turquoise matrix is also used now for medium class jewelry, the cutting including both the stone and its matrix. The turquoise in a dark-brown matrix is much fancied for this purpose, as the mottling of brown in the blue produces a very rich effect. The matrix of gems from some American mines is flinty, and both the gem and the matrix are very hard which affords possibilities of a high polish, but as the flint sometimes penetrates the turquoise it is apt to break it.

Occidental turquoise, formerly used extensively, is odontolite, made from fossil bone, coloured by a phosphate of iron; it is still mined to a small extent in the vicinity of Simor, Lower Languedoc, France. This western "turquoise" loses its colour in artificial light, and, when heated, gives off an offensive odour caused by the decomposition of animal matter. Its weight is lighter than that of turquoise, and it does not give a blue colour, with ammonia, when dissolved in hydrochloric acid, like the genuine.

The conditions peculiar to the demand for turquoise at present in America are like those affecting opals; the very choicest specimens are highly prized and readily sold, while the average specimens are considered with indifference.

CHAPTER XIV

CAT'S-EYE

CAT'S-EYE is a well established term in the trade in precious stones, and more than one mineral which exhibits *chatoyancy*—a French word signifying a changeable, undulating lustre, like the eye of a cat in the dark—is termed, and sold as "cat's-eye."

The true cat's-eye is cymophane, a variety of chrysoberyl, a mineral resembling beryl in containing the element glucinum (beryllium), but otherwise distinct. Chrysoberyl is devoid of silica, which beryl possesses, and is, theoretically, composed of glucina, 19.8 and alumina, 80.2. Jewellers variously call chrysoberyl "cat's-eye," "Oriental cat's-eye," or "Ceylonese cat's-eye." Besides its principal components, chrysoberyl frequently contains impurities such as iron and chromium oxides. Chrysoberyl is very hard—8.5, being third in Mohs's scale to the diamond, and when cut is susceptible of a high polish. Heavier than the diamond, the specific gravity of chrysoberyl ranges from 3.5 to 3.8. Chrysoberyl crystallises in the rhombic system and commonly appears in complicated twin crystals. This peculiar mineral has no distinct cleavage, but has a conchoidal fracture; it is brittle; acids will not attack it; it is infusible before the blowpipe; it can be electrified positively, by friction, and will remain charged for several hours. Lustre vitreous to slightly greasy. Chrysoberyl is transparent to opaque, but is only transparent when cut and polished; it is doubly, but not strongly, refractive. The limited range of colour in Brazilian specimens is from pale yellowish-green to golden yellow and brownish-yellow. Crystals from the Ural Mountains vary from an intense green to grass-green or emerald-green—the latter variety is alexandrite.

The distinction of cymophane from ordinary chrysoberyl is its chatoyancy, which appears as a milky-white, bluish or greenish-white, or, more rarely, golden-yellow sheen which follows every movement of the stone; this characteristic is most strongly developed by cutting the stone convex, and therefore cat's-eye is cut *en cabochon*. A silvery line or streak of light extends across the curved surface and is most strongly defined in a strong light, while its boundaries are sharpest in small stones. The effect of the chatoyancy is in great part due to the judicious work of the lapidary, and usually the greatest possible effect is produced by the greatest curvature of the surface. Chatoyancy appears only in the cloudy chrysoberyl, and the cloudiness is due to thousands of microscopically small cavities within the stone. The influence of the whims and preferences of royalty on the popularity of gems was remarkably illustrated by the sudden favour with which chrysoberyl cat's-eye was invested, when His Royal Highness, the Duke of Connaught, gave his *fiancée* a ring set with this stone, which vastly increased the demand for it and caused a corresponding rise in price.

The Minas Novas district in the northern part of the state of Minas Geraes, Brazil, is the most prolific producer of chrysoberyl of the finest colours; most of the specimens are chatoyant. The mineral in this locality occurs associated with rock crystal, amethyst, red quartz, green tourmaline, yellowish-red (vinegar) spinel, garnet, euclase and white and blue topaz. Chrysoberyl is erroneously indentified with, and termed, chrysolite by the Brazilians, and this error is prevalent in the trade in precious stones and jewelry, almost everywhere. The usual tests, the scale of hardness especially, will promptly differentiate chrysolite. The source of supply of cymophane and non-chatoyant chrysoberyl second in importance to Brazil, is the island of Ceylon. The cat's-eye record for size was long held by a Ceylonese specimen, and, until the year 1815, this was a jewel in the crown of the King of Kandy. The weight of the Ceylon stones ranges from one to one hundred carats; they are found in company with sapphires in gem-gravels, chiefly in the Suffragan district and the vicinity of Matura in the south of the island. To a small extent, chatoyant chrysoberyl is mined in the Ural Mountains of Siberia.

Among the numerous minerals which when fibrous, or cut across the cleavage and convex, will exhibit the opalescent ray resembling the contracted pupil of the eye of a cat, are beryl, corundum, crocidolite, dumortierite, quartz, filled with acicular crystals or fibrous minerals, such as actinolite, byssolite, or hornblende; hypersthene, enstatite, bronzite, aragonite, gypsum, labradorite, limonite, and hematite. These may be opaque, translucent, or transparent and of any colour or colours. Perhaps the commonest of these minerals is the quartz cat's-eye, which falls far short of rivalling the brilliancy and soft colouring of cymophane. The shades of this variety of quartz are greenish, yellowish-grey, and brown. Simple tests will distinguish this mineral from cymophane, as its hardness is but 6 to 7 and its specific gravity, 2.6. This quartz melts with soda to a clear glass, is soluble in hydrofluoric acid, and is not dichroic; its chief components are silicon and oxygen. Cut *en cabochon*, a band of light appears across the parallel fibres of asbestos which the quartz contains.

Tiger-eye, in the trade, is considered separately from cat's-eye, but as chatoyancy is its chief characteristic, it may as well be included here and, as its present commercial value is low and the demand for it is small, it can be summarily described and dismissed. The proper term for the mineral known as "tiger-eye" is crocidolite, a name derived from the Greek and meaning "woof," in allusion to its fibrous structure. Crocidolite is a fibrous asbestos-like mineral. Its colours are gold-yellow, ranging to yellowish-brown, indigo to greenish-blue, leek-green and a dull red. The blue is usually distinguished as "hawk's-eye." Crocidolite contains a siliceous base, usually a ferruginous quartz, and when cut highly convex with the longer diameter of the oval at right angles to the direction of the fibres, the cat's-eye ray is strongly apparent. Crocidolite contains: silica, 51; iron oxides, 34; soda, 7; magnesia, 2; water, 3. Hardness 4 to 7 and specific gravity 3.26. The best specimens are found in the Orange River region and Griqualand, South Africa.

Tiger-eye is well adapted to, and has been largely used for carving cameos and intaglios; it was very popular from about the year 1880 to 1890 in the United States.

The stones distinguished as chatoyant sometimes include alexandrite, a variety of chrysoberyl, strongly dichroic and sometimes trichroic. Mr. Edwin W. Streeter, in his book *Precious Stones and Gems*, states that he has seen specimens of alexandrite with a perfect cat's-eye line, yet subject to the change of colour by artificial light characteristic of this mineral. To display the ray, the stone is of course cut convex instead of with six facets. This stone was discovered in the Ural Mountains, Siberia, in the year 1830, on an anniversary of the birthday of the Czar Alexander II., of Russia, for whom it was named. Alexandrite has marked hues of red and green, the national colours of Russia; by daylight it shows a bright or deep olive-green colour, but in artificial light a soft columbine red or raspberry red or raspberry tint. One description of this gem includes the phrase "it is an emerald by day and an amethyst by night." Subsequent to the discovery of alexandrite in the Urals, the same gem mineral, but of a better and more workable quality, was discovered in the island of Ceylon, which is the present principal source of supply.

CHAPTER XV

CHRYSOPRASE

CHRYSOPRASE is the chief of two varieties of hornstone which are cut as ornamental stones, the other being wood-stone or silicified wood, such as is obtained from the petrified forest known as Chalcedony Park, in Arizona, and which occurs abundantly in various mountainous localities in the western United States. Hornstone is an old mining term and is not used by lapidaries. It is a fine-grained, very compact, variety of quartz, of a granular consistency.

The name chrysoprase is derived from two Greek words, meaning golden leek, and describes the colour of the stone. The ancients ascribed to it the virtues of the emerald, though in a lesser degree. They believed it lost its colour when in contact with poison, and was a cordial and stimulant.

A characteristic of chrysoprase is its splintery fracture; the sharp edges of fragments verging on translucency. The approved tints of chrysoprase are leek and apple green, although the blue, golden-green, and other yellowish tints are occasionally used. The colours remain steadfast in artificial light. The colour owes its presence to about one per cent. of nickel, probably in the form of a hydrated silicate; the loss of water through heating the stone but moderately, causes it to pale gradually, until it ends in a total loss of colour. A long exposure to the direct rays of the sun will produce a like effect, but the cause will be the strong light and not the heat. The brittleness of chrysoprase presents difficulties to the lapidary; it is usually cut *en cabochon*, or else with a plane surface bordered with one or two courses of facets. Although its intrinsic value is less than it was formerly, chrysoprase is one of the most valuable varieties of quartz in the ornamental stone field, and is highly esteemed among the semi-precious stones.

Chrysoprase occurs in plates and veins, usually locked in serpentine, and its most ancient and common source is a district south of Breslau in the province of Silesia, Germany. According to an account published in 1805, a vein of chrysoprase three (German) miles long was discovered in 1740 by a Prussian officer. The real discovery probably long preceded this, because chrysoprase, used decoratively, has existed in the Wenzel Chapel, Prague, since the fourteenth century. The leek-green stone is found in a few other unimportant localities in Europe, also India, in the Ural Mountains, Siberia, and it occurs in various places in North America; one is at Nickel Mountain near Riddle, Douglas County, Oregon, but the most important mines are those of the Himalaya Mining Company, about eight miles from Visalia in Tulare County, California.

Frederick the Great of Prussia highly favoured and evinced a great interest in this beautiful stone; possibly this was to some extent because it originated in Silesia, which became his conquered territory in 1745, after his second Silesian war. Frederick had two famous tables made of chrysoprase, and had it utilised in mosaics. Basking in the sunlight of royal favour, chrysoprase grew in popularity, which its native merits have always, to a considerable degree, sustained.

A charming Roumanian legend ascribes the discovery of chrysoprase in the rocky bed of the Riul Doamnei, a beautiful stream, to a Princess Trina, who, to succour her people in time of dire famine, stripped herself of all her possessions but a pitiful last piece of jewelry, a golden lizard with green eyes of chrysoprase, given to the princess on her wedding day by her deceased mother. A wizard admonished the princess never to part with the lizard, because it would some day bring untold riches, and besides that, whoever possessed any leek-green chrysoprase would, in time of great distress, understand the language of animals. Reduced to the verge of selling her last treasure by the unbearable sight of the sufferings of the children of her starving people, the good Princess Trina was weeping and praying at a window, when a tiny lizard with glittering green eyes darted into the room, and, in a silver voice and lacertilian language, which the princess by virtue of her talisman understood perfectly, said: "Help shall arise for thee out of a river: Only seek."

Thus admonished the princess wandered through the stony bed of one river after another wearing out her eyes, her strength, and her soul, in the search; until, when about to succumb to exhaustion, she discovered a vast treasure of chrysoprase, thus ending the famine and inaugurating an unprecedented reign of prosperity for her beloved people.

Besides the remarkable understanding of the lizard's speech by the princess, another miraculous occurrence is connected with this discovery: from that day to this, the waters of the Riul Doamnei have remained a leek-green, as can be easily proved to any one visiting the place.

CHAPTER XVI

JADE

JADE is a verdant mineral known to man for ages, and used for personal ornaments, weapons, implements, art objects, and applied to interior decoration. The word emerald, so frequently appearing in ancient writings, is believed to have sometimes meant jade—an opaque to translucent mineral—and unlike the emerald in anything, excepting a slight resemblance in colour. The word "jade" is now a generic term applied to various mineral substances, as chloro-melanite, or jadeite, nephrite, saussurite, pseudo-nephrite; these minerals are characterised by toughness, compactness of texture, and a colour range from cream white to dark green and nearly black. Although appearing in the trade in precious stones and jewelry, in the art objects of every land, and although extensively imitated—sometimes in a fashion, however, that could deceive no one—"jade" is nowhere prized and appreciated so much as in the Chinese Empire; and wherever on the globe adventurous Chinese roam or locate it is always found as one of their most cherished possessions. Properly the term "jade" includes but two minerals; nephrite and jadeite. Nephrite is *Nephrus amorphous* of the order Chalicinea, according to Dana's system of mineralogy. The name is from a Greek word meaning a kidney; the ancient Greeks believing this mineral to possess the virtue of a specific remedy for all diseases of the kidneys, as, indeed, the Chinese believe now, and have for centuries. Jade is massive, of fine granular or impalpable substance; hardness, 6.5; specific gravity, 2.96 to 3.1; lustre, vitreous; streak, white; colour, leek-green, passing into blue, grey, and white; translucent to sub-translucent; fracture, coarse and splintery. An average specimen contains silica, 50; magnesia, 31; alumina, 10; oxide of iron, 5.5; and nearly three per cent. of water, with a tinge of chrome oxide. Jade is infusible before the blowpipe, but becomes white; with borax it forms clear glass.

Jadeite is a tough, fibrous foliated, to closely compact, mineral, grouped with the pyroxenes; hardness, 6.5 to 7; specific gravity, 3.33 to 3.35. Jadeite will fuse readily before the blowpipe to a transparent glass containing bubbles or blisters. A variety that is dark green verging on black is termed chloromelanite. Weapons and ornaments carved in jadeite in prehistoric times are found on every continent. But few of the localities from whence the mineral came that supplied raw material for these unnamed artisans and artists, are known; the most important is in the vicinity of Mogoung in Upper Burma, where it occurs in boulders embedded in a reddish-yellow clay in river valleys. The jadeite miners crack the boulders by heating, and the pieces found of merchantable quality are either sawed into the required shapes by slender steel saws, kept tense by bamboo bows, or sold as found to traders who come in caravans from China. The mineral here found is thus distributed throughout the Chinese Empire. Jadeite of milk-white colour is most highly prized and that with bright green spots is next in favour. Dr. Max Bauer states that he saw a piece of less than three cubic feet which sold for $50,000.

Nephrite occurs in gneiss and amphibole schists in the Karakash Valley in the Kuen Lun Mountains, Turkestan, and this is now an important source of supply; these mines have been worked for more than two thousand years. Nephrite is found in eastern Siberia, Silesia, Germany, and in New Zealand. Both nephrite and jadeite, carved into weapons and ornaments, have been found in all the Americas; the occurrence of nephrite in Alaska has been well established, and it is a possibility that much of the carved material found far south of Alaska originated there.

The Chinese name for jade is "Yu," or "Yu-Shih" (Yu stone), and the Chinese do not seem to distinguish between jadeite and nephrite. In the western world jade is used but to a limited extent for jewelry, excepting as an artistic fancy or fad, by those who have visited the Orient, or become interested in it through visiting the "Chinatown" colonies of the immigrant Cantonese in American cities. A demand for jade bracelets as souvenirs of visits has grown up, these Oriental ornaments being especially appreciated by the artistic. Outside the realm of jewelry, very high prices are paid in Europe and the United States by connoisseurs and collectors for beautiful examples of Chinese art, not for the intrinsic value of the mineral, but because of the wondrous workmanship displayed by the patient and skilful Chinese artisans.

CHAPTER XVII

MOONSTONE

MOONSTONES have a soft attractiveness that is in contrast with the flashing angles of the majority of precious stones. They are usually cut *en cabochon* or sometimes turned in the form of balls, and, as the stone is reputed to be potent in providing its possessor with good fortune, these chatoyant spheres are in favour as lucky charms. The superstitions regarding gems in medieval times included one that was quite general, that a moonstone held in the mouth would stimulate and refresh the memory. If the moonstone really possesses such efficacy, it should be a modern specific for witnesses in courts of justice, such as corporation officers whose books have been burned, or otherwise illegally disposed of, and bankrupts who cannot remember what disposition was made of their assets. Among the beliefs held of this stone, was one that it would cure epilepsy, a faith still retained by the French peasants of the Basque province. Another belief was that during the waxing of the moon it was an efficacious love charm; while during the moon's waning it would enable its wearer to foretell future events. If there is any basis in fact for this belief, it should be the favourite gem of tipsters of the race tracks and stock market.

A sort of cousin-german of the moonstone is the sunstone, which however is a far less important luminary in the firmament of gems. Although various minerals may be termed "moonstones," the true moonstone is the opalescent variety of orthoclase-feldspar, also bearing other names, but usually identified by the name adularia—a name which it derives from Mount Adula, one of the highest peaks of St. Gothard in the Alps, where it is found. The Greeks called it *Aphroseline*, signifying the splendour of the moon. The Romans called it *Lunaris*. A transparent, fibrous, lustrous gypsum, found in England, selenite, which derives its name from its soft lustre, suggestive of moonshine, and literally signifying "moonstone," may be merely mentioned here, but this soft substance is entitled to no place in a list of even the semi-precious stones.

Moonstone, according to the mineralogical concepts of the United States National Museum, is a transparent albite having a chatoyant reflection resembling that of a cat's-eye, or an opaque pearly white albite having a bluish opalescence. Albite occurs in opaque to transparent masses and in triclinic crystals having a dual cleavage in different directions, one of which is highly perfect; hardness, 6; specific gravity, 2.62; lustre vitreous, sometimes pearly on a cleavage surface; colours, white, bluish, greyish, reddish, greenish, and green, with, occasionally, a bluish chatoyancy or play of colour. One hundred parts of albite contain: silica, 68.7; alumina, 19.5; soda, 11.8.

Albite is a constituent of many crystalline rocks, and frequently replaces feldspar as a constituent of granite, of syenite, and of greenstone; sometimes it is associated with feldspar and dolomite. Common occurrences are in veins or cavities in granite or granitoid rocks, which are also sometimes repositories of fine crystals of other gem minerals, such as beryl, tourmaline, and smoky quartz.

The moonstone of commerce comes chiefly from Ceylon, where it is found in pieces several inches in diameter resulting from the decomposition of a porphyritic rock. Ceylon moonstone is sometimes erroneously termed "Ceylon opal." Albite is found at Mineral Hill, near Media, Delaware County, Pennsylvania; in Allen's Mica Mine, Amelia Court House, Virginia; and other localities in North America.

The term sunstone, or heliolite, is applied to aventurine kinds of oglioclase, one of the feldspars; these are of a greyish white to reddish gray colour with internal yellowish or reddish reflections, proceeding from disseminated crystals or flakes of iron oxide. Sunstone is found at Lyme, Connecticut, among other American localities. Its use in jewelry is now very limited; it is not costly, and artificial "sunstone" or "goldstone," made of glass, containing sparkling particles of metal, is often preferred to the genuine.

CHAPTER XVIII

PERIDOT

HYBRIDS are foreign to mineralogy, but there is no precious stone so difficult to specifically determine as chrysolite, because of the confusion regarding it in the minds of those engaged in the commerce of precious stones.

Mineralogists generalise the varieties of chrysolite under the common term "olivine." To American jewellers it is perhaps most commonly known as peridot. With the usual indifference to mineralogical distinctions of the average jeweller, it is possible that more green garnets than chrysolite are sold under the name olivine. W. R. Cattelle, in his book, *Precious Stones* writes:

> The distinction between varieties is practically one of colour only. For many years lapidaries were in the habit of calling the chrysoberyl "Oriental chrysolite," and in consequence the two stones have been confused, though the chrysolite is much the softer stone and usually shows marked differences in colour and lustre.

At present it is customary to call those which incline most to yellow "chrysolite"; the yellowish green, resembling a light tourmaline with a dash of yellow, is known by the name "peridot," given to it by the French jewellers; and "olivine" is the name associated with the brighter yellowish emerald-green variety, although originally the yellow to olive-green stones were known by that name.

Few olivines are sold as such. The beautiful bright yellowish-green stones known here as olivines, are generally demantoids, Russian green garnets, of about the same hardness; these are rarely found large enough to cut to gems of over one half to three quarters of a carat.

Olivine crystallises in the orthorhombic system; also occurring massive; compact or granular; usually in embedded grains; hardness, 6.5 to 7; specific gravity, 3.33 to 3.44; cleavage, distinct; fracture, conchoidal; brittle; lustre, vitreous; colour, typical, olive green; brownish, greyish red and black. It is strongly doubly refractive with marked dichroism in some specimens; peridot showing straw-green and a green image. Gem kinds and their colours are chrysolite, yellowish green; peridot or "evening emerald," olive pistachio, or leek-green colour, of a hue more subdued than the emerald—green beryl. The approved tint of peridot resembles that revealed by looking through a delicate translucent green leaf. Hyalosiderite, "Job's tears," is a highly ferruginous variety; specific gravity attaining 3.57; colour, a rich olive green.

Olivine is a frequently occurring constituent of some eruptive rocks, is also found in granular limestone and dolomite, and in several schists and ore deposits. Chemically, olivine—a sample specimen—is composed of, approximately, silica, 41; magnesia, 50; iron oxide, 9.

Olivine is a constituent of meteorites. The sources of supply of this somewhat puzzling mineral are characteristically doubtful. Dr. George Frederic Kunz is quoted as saying that our modern supply of chrysolite is taken out of old jewelry. The large transparent pieces of chrysolite used for gem purposes are reported to originate in the Levant, Burma, Ceylon, Egypt, and Brazil. Recently a limited supply has come into the market from upper Egypt near the Red Sea—perhaps an ancient source. The chrysolite of the Bible may have been topaz. Small chrysolites—"Job's tears"—of good quality are found in the sand with pyrope garnet in Arizona and New Mexico.

CHAPTER XIX

KUNZITE

KUNZITE is a comparatively new transparent gem discovered in America about 1903; it is a lilac-coloured spodumene, which, upon the suggestion of the mineralogist Charles Baskerville, was named kunzite, in honour of Dr. George Frederic Kunz, because of his services to the scientific world in the gem branch of mineralogy. The honour accorded Dr. Kunz by mineralogists in accepting the name is enhanced because of the beauty of this new gem mineral. The first crystals of this unaltered lilac-coloured spodumene were discovered a mile and a half northeast from Pala, San Diego County, California. The vicinity of this discovery was already of great interest to students of gem minerals because but fifty feet away from the spot is a famous deposit of tourmaline from which specimen crystals remarkable for the unusually large size and great beauty have been taken, while half a mile away is a celebrated rubellite and lepidolite locality. The spodumene crystals found near Pala are of extraordinary size, one weighing thirty-one ounces, troy; the dimensions of this crystal were 18 x 8 x 3 centimetres.

Kunzite has a considerable range of tints which include shades characterised as: deep rosy lilac, rich deep pink purple, and delicate pink amethystine; this and the lighter lilac shades are the typical tints. The finest specimens we have seen have a bright lustre and perfect transparency. These lilac-spodumene crystals occurred in a ledge which was traced for twelve hundred feet along the top of a ridge. The rock is a coarse decomposed granite, which might be termed pegmatite, with the feldspar much kaolinised and reduced to a "red dirt," and showing many large quartz crystals, some of them weighing 150 pounds, but not clear.

Other coloured crystals of spodumene which approach in colour and quality the standard specimens obtained near Pala have been found at Meridian, California, but these are smaller than those found at Pala; the Meridian specimens more nearly resemble the occasional specimens of unaltered spodumene found near Branchville, Connecticut. The Meridian crystals were at first supposed to be tourmaline, but were identified by Dr. Kunz; many of these crystals were ruined by lapidaries who unsuccessfully tried to cut them, as the very highly facile cleavage of spodumene caused the mineral to flake.

Kunzite is entirely distinct from the green variety of spodumene (hiddenite), the beautiful gem mineral found at Stony Point, Alexandra County, North Carolina, and from the transparent yellow variety reported by a mineralogist named Pisani to have been found in Brazil, and, since its discovery, produced in sufficient quantity to come into use as gems.

Spodumene—it is also sometimes called triphane—in its general characteristics is a member of the pyroxene group, and is the only gem mineral, besides lepidolite and tourmaline, which contains a considerable proportion of lithium. The chemical composition of spodumene is: silica, 64.5; alumina, 27.4; and lithia, 8.4. Spodumene is fusible before the blowpipe; its hardness is 6½ to 7; specific gravity, 3.1-3.2; lustre, vitreous. Spodumene is commonly white or grey, and because of that it was named, the word spodumene being derived from the Greek *spodios,* meaning ash-coloured. Most of the spodumene found is opaque, only the gem quality being translucent to transparent. Spodumene crystallises in the monoclinic system, and crystals have been found four feet long.

Until the discovery of kunzite the use of spodumene as a gem was limited to the emerald-green hiddenite, named after its discoverer, W. E. Hidden. This variety occurs in thin crystals with tints ranging from colourless to yellow and to an emerald green. Five carats is about the maximum weight of cut hiddenite gems; they are cut into step or table stones to make the most of their dichroism, and to avoid the possibility of splitting because of their unusually high degree of prismatic cleavage.

The Brazilian spodumene, the yellow, was originally identified as chrysoberyl, and it is used in jewelry as the last named metal is; scientific tests will easily distinguish these two minerals the one from the other. Some spodumene of a beautiful blue colour has also been found in Brazil, near Diamantina.

Kunzite, almost 7 in hardness, is transparent and pleochroic. Viewed transversely some representative crystals were faintly pink; longitudinally they presented a rich pale lavender colour, approaching amethystine. A characteristic of kunzite crystals is a peculiar etching, apparently effected with solvents. A number of scientific tests have revealed in kunzite a remarkable phosphorescence, not possessed by other varieties of spodumene similarly tested, and its illuminant powers, excited by its bombardment with Röntgen rays, and also by the proximity of a few milligrammes of radium bromide, mark this mineral as unique and of unusual interest to scientists, in addition to its value as a recruit to the first rank of semiprecious stones.

In a description of experiments made upon kunzite Sir William Crookes writes:

> But the most interesting thing to me is the effect of radium on it. A few milligrammes of radium bromide brought near the piece of kunzite makes it glow with a fine yellowish light, which does not cease immediately on removal of the radium, but persists for several seconds.
>
> I have found some diamonds phosphoresce brightly under the influence of radium, and have been searching for a mineral which is equally sensitive. I think this lilac variety of spodumene runs the diamond very close, if it does not surpass it sometimes.

The luminosity of kunzite, in response to the artificial conditions already known to arouse it, is thus summed up, in a sentence, by Dr. Kunz:

> In a word, kunzite responds to radium, actinium, Röntgen and ultra-violet rays; it is thermo-luminescent and pyro-electric. Becomes radescent when mixed in powdered form with radium; becomes incandescent when this mixture is slightly heated, and crystals or gems become beautifully phosphorescent for quite a time by passing a faradic current through it, or if held between the poles of a Holtz machine.

The sole drawback at present to the increasing appreciation of kunzite is that the supply, according to reports in the jewelry trade in New York City, is unequal to the increasing demand. In 1907, according to reports of the United States Geological Survey, about 126 pounds of gem spodumene, selected material, was obtained from the California gem region, but not all of this was the variety kunzite. Albert Dabren, a mining engineer, of Madagascar, has reported that gem kunzite has been found there.

CHAPTER XX

TOURMALINES

A STONE of many colours is tourmaline; it was introduced into Europe from India in 1703 and its name is adapted from *turmali*, its Cingalese name. Tourmaline is a widely distributed mineral, and its transparent coloured varieties, used as gem stones, have attained a considerable popularity. The vogue of the tourmaline has increased since it was discovered in 1820 on Mount Mica near Paris, Maine. The tourmaline has also been found in Massachusetts, California, and New York State. Its principal sources are Ceylon, Burma, Brazil, and the Ural Mountains, Siberia; it is also found in Moravia, Sweden, and the Isle of Elba. Tourmaline occurs in granite, particularly the albitic varieties, schists, and dolomite. Crystallisation of the tourmaline is rhombohedral, hemimorphic, and the prisms have three, six, nine, or twelve sides. In hardness it is equal to quartz and approaches topaz, being 7 to 7.5. Its lustre

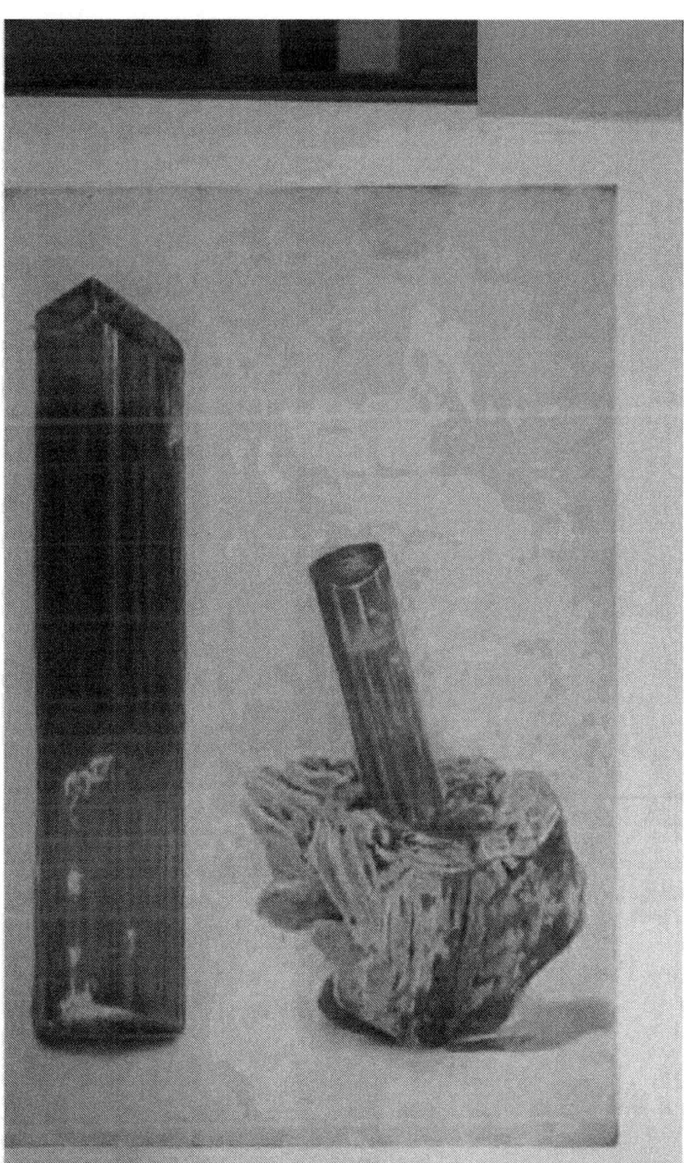

GREEN AND PINK TOURMALINE, MESA GRANDE, CALA.; OWNED BY HARVARD UNIVERSITY.
Courtesy of A. H. Petereit

PINK TOURMALINE IN ALBITE WITH LEPIDOLITE, MESA GRANDE, CALA.
Theo. Henninger

is vitreous, it ranges from transparent to opaque, and is doubly refractive to a high degree. Its cleavage is perfect on the basal plane, breaking with uneven fractures. Its specific gravity is from 2.94 to 3.15.

Tourmaline is one of the most dichroic stones, and individual specimens vary more from others in composition and proportion than is the case in almost any other mineral. In colour, black shading to light brown is the commonest; but blue, green, red, and pink are usually desired. Some of the shades are very rich; and richness, rather than brilliancy, is the quality which appeals to the artistic eye of the connoisseur. Curious specimens have shown internal shades of red and external of green, while others differ in colour toward the extremities. Dana distinguishes varieties as follows: rubellite, shades of red, frequently transparent (two of the finest known specimens of this variety are in the British national collection in the Natural History Museum at South Kensington, England); indicolite, indigo blue—Berlin blue, the Brazilian sapphire of jewellers; Brazilian emerald; chrysolite (or peridot), green and transparent; peridot of Ceylon, honey-yellow; achroite, colourless; aphrizite, black; and columnar and black, without cleavage or trace of fibrous texture.

Tourmaline heated, like some other minerals in which one termination differs in form from the other, develops electricity, with the effect of making of the ends positive and negative poles. Sections of tourmaline crystals cut parallel to the axis have the property of polarising light. Tourmaline can be fused under the blowpipe to a spongy enamel; it melts with borax to transparent glass. Tourmaline is cut step and brilliant.

Twin-coloured tourmaline is strongly doubly refractive; green shows yellow and greenish blue; yellowish green, yellow and green; reddish brown, light and dark brown; red, pink and dark red; blue, light and dark blue. The green tends toward blue while the blue has a greenish tendency. Some brown tourmalines have mixed colours.

In considering shades when selecting tourmalines, a medium bright green is better than the lighter or that which appears blackish. The pink should be deep and clean, ruby-like. A rich amber brown is most desirable of the brown shades. Red tourmaline is occasionally so like the ruby that it might deceive any but the expert and his recourse to a scientific test; the hardness of the ruby would of course decide it. In its two-colour character, the tourmaline resembles the ruby but surpasses it; the colour of the tourmaline is not so deep nor is it so lustrous as the ruby's, but it is frequently more transparent. While some red tourmaline resembles spinel, the latter is singly refractive and has a yellow tint. Red topaz is harder and of greater specific gravity than red tourmaline. The two colours of the topaz are red and yellow while the tourmaline's are rose and dark red. Sapphire is harder than tourmaline and clear blue, while tourmaline is greenish blue. Aquamarine is a water blue and is harder than tourmaline, but is of a lower specific gravity. The several other colour varieties of tourmaline bear sometimes a strong resemblance to other stones, but are easily distinguished by the expert, usually without further test than the employment of the dichroiscope. Tourmaline has sometimes been confounded with some of the fine green diopsides found in New York State.

Digging for tourmalines, at least in one locality, offers the fascination that, in some form, seems always present in the mineral industries. One of the earlier sources of supply of tourmalines was Burma, and an interesting description of some of the phases of the quest for tourmalines was written by Mr. C. S. George, deputy commissioner for the Ruby Mine District, Burma, for the London *Tribune*. Tourmaline, as found there, is in separate crystals in the interstices of granite rock, and men with no capital can mine here and do, in a desultory manner, on the chance of finding more or less valuable bits by digging down a distance of ten feet or less. This was the method of mining at the original ruby diggings at Kathe. The more modern method is that of sinking a vertical shaft four or five feet square. Custom allows the proprietor of the shaft to extend his workings underground anywhere to a radius of five fathoms from the centre of the shaft.

A writer—Mr. C. S. George referred to above —in the *Jeweller's Circular Weekly* states:

> The vein is formed by a vein of white hard granite rock, in the interstices of which the tourmaline is found, at times adhering loosely to the rock, at others lying separate in the loose yellowish earth that is found with granite. When a vein is once found it is followed up as far as possible, subject to the five fathom limit. What, however, makes the mining so exciting and at the same time keeps the industry fluctuating is that the tourmaline crystals

are found only intermittently in the vein. One may get several in the length of one yard, and then they will unaccountably cease. Directly one man strikes a vein yielding crystals every one who can commences digging along the line of the vein, but it is all a toss-up as to whether, when the vein is reached, there will be tourmaline therein. Adjoining shafts give absolutely different results, and it is calculated that at least two thirds of the shafts sunk yield nothing at all, while only an occasional one is at all rich.

Of the sixty-two shafts at the time of Mr. George's visit only three were yielding, and of these only one had traces of the best quality stone. The veins are fairly deep down, none having ever been reached at a lesser depth than nine fathoms, while an ordinary depth is forty or fifty cubits. When the "vein" takes a downward direction it is followed as far as possible, but that is rarely over about sixty cubits, for at that depth the foulness of the air puts the lamps out.

All the material dug out from the inside shaft is pulled up to the surface in small buckets, all worked by enormously long pivoted bamboos weighted with a counterpoise, and the tourmaline is sorted out of hand, the granitic fragments being piled in a wall around the mouth of the shaft.

The folk-lore of tourmaline tells us that both the introduction of this beautiful and multiphase mineral to the knowledge and appreciation of mankind, and its discovery in America, were due to children. Soon after the year 1700, some children in Holland were playing in a court-yard on a summer day with a few bright-coloured stones indifferently given to them by some lapidaries, who evidently had not classified, or invested them with any particular value or significance. The children's keenness of observation revealed that when their bright playthings became heated by the sun's rays, they attracted and held ashes and straws. The children appealed to their parents for enlightenment as to the cause of this mysterious property; but they were unable to explain or to identify the stones, giving them, however, the name of *aschentreckers* or ash-drawers, which for a long time clung to these tourmalines.

The story of the tourmaline in the western hemisphere is an object-lesson for those adults who have no indulgence for the scientific enterprise of the young, or faith in the possibility of valuable results from their immature investigation. The principal source of the best American tourmalines is a mine on Mount Mica at Paris, Maine. Gem tourmalines were discovered on Mount Mica on an autumn day in 1820 by two boys, Elijah L. Hamlin and Ezekiel Holmes, amateur mineralogists. When nearing home from a fatiguing local prospecting expedition, they discovered some gleaming green substance at the root of a tree, and investigation rewarded them with a fine green tourmaline. A snowstorm prevented a further search, but the following spring they returned to their "claim" and secured a number of fine crystals. Tourmalines from Mount Mica are found in pockets in pegmatitic granite, overlaid by mica schist, which has since to some extent been stripped off to facilitate this interesting mineral industry. Black tourmaline, muscovite, and lepidolite are found in this Pine Tree State treasure house. More than fifty thousand dollars' worth of tourmalines have been extracted from the mine resulting from this boyish discovery. While this sum of money is not great in comparison with the financial results of many mineral industries, the output has included very many specimens of rare beauty that have enriched the collections of royalty, wealthy private connoisseurs of precious stones, and of great public museums and educational institutions.

The strong dichroism of the tourmaline and its variety of colour composition and other remarkable properties make it one of the most interesting minerals in Nature's storehouse, and led Ruskin to write in his *Ethics of the Dust*, in a fanciful effort to describe its harlequin composition:

> A little of everything; there's always flint and clay and magnesia in it; and the black is iron according to its fancy; and there's boracic acid, if you know what that is, and if you don't, I cannot tell you to-day, and it does n't signify; and there's potash and soda; and on the whole, the chemistry of it is more like a mediæval doctor's prescription than the making of a respectable mineral.

CHAPTER XXI

AMBER

ALTHOUGH the ornamental uses of amber are to a great extent outside the realm of personal adornment, its conversion into beads, for necklaces especially, is of such ancient origin, and these ornaments have always been so favoured, that this fossil vegetable resin is, like the pearl and coral, included in the realm of gems which are, with these exceptions, and the diamond, which is carbon, purely mineral. Like the pearl and coral, amber is identified in the popular conception with the sea, from whence a small proportion of the amber acquired by man has been derived.

To use the words of Dr. Max Bauer: "This material, so much used for personal ornaments, is not strictly speaking a mineral at all, being of vegetable origin, and consisting of the more or less considerably altered resin of extinct trees. It resembles minerals in its occurrence in the beds of the earth's crust, and for that reason may be considered, like other varieties of fossil resin, of which it is the most important, as an appendix to minerals."

Archæological discoveries reveal that amber was known to and favoured by prehistoric peoples, such as the Egyptians and cave-dwellers of Switzerland. Amber is believed to have been taken from the Baltic by the seafaring Phœnicians, and the old Greeks called it *elektron*, from whence comes our modern word electricity.

True amber—*Succinum electrum* (Dana)—the succinite of mineralogists, is the resin of a coniferous tree which was of the vegetable life of the Miocene age of the Tertiary period in geology. The late Professor Goeppert, of Breslau christened the principal amber-yielding tree the *Pinites succinifer*. The vegetable origin of amber has not been definitely established in science, but one of the evidences that it was a flowing vegetable resin, that is accepted as indisputable, is the oft-occurring presence in amber of insects, or parts of them, which must have been caught and imprisoned when the fresh resin was fluent. Wherever amber is found in the earth, it is in association with brown-coal or lignite.

Amber, or succinite, then, is a fossil resin

occurring in irregular masses with no cleavage and having a conchoidal fracture. Colour yellow, some specimens reddish, brownish, whitish, or cloudy and occasionally fluorescent, with a blue or green tinge; hardness, 2 to 2.5; specific gravity, 1.05 to 1.09; brittle; lustre, resinous to waxy; transparent to opaque; negatively electrified by friction. Amber is inflammable with a rich yellow flame and it emits an aromatic odour; heated to 150 degrees C. it softens, and melts at about 250 degrees C. giving off dense white pungent fumes. In alcohol it is soluble. The chemical constituents of amber, in one hundred parts are: carbon 78.96, hydrogen 10.51, oxygen 10.52.

Amber is found on the Baltic, Adriatic, and Sicilian coasts; in France, China, India, and in North America.

Always within man's memory or knowledge, nodules of amber have been cast up on the shores of the Baltic Sea, especially along the Prussian coast, and their collection and sale has afforded a livelihood for the local inhabitants. This is called "sea stone," or "sea amber," and it is usually uniform and, being uncontaminated by associated substances, is superior in quality to that which is mined.

This flotsam amber is often entangled in sea-weed and this—called "scoop stone"—is collected in nets. In marshy spots, mounted men, called "amber riders," follow the ebbing tide and profitably search for the fossil resin thus exposed. The weight of amber being about the same as sea-water, agitation of the water containing it is sufficiently effective for its flotation. About 1860, it being evident to geologists that the sea-amber came from the strata underneath, it was sought on the adjacent terra firma by modern mining methods, and the operations have resulted in an established successful industry.

The most highly prized amber comes from Sicily. Professor Oliver Cummings Farrington, in his book *Gems and Gem Minerals*, states that eight hundred dollars have been paid for pieces of Sicilian amber no larger than walnuts. The Sicilian amber reveals a varied colour display including blood-red and chrysolite-green, which are often fluorescent, glowing internally with a light of different colour from the exterior. The advantages of amber, despite its softness, include its remarkable durability.

CHAPTER XXII

BLOODSTONE

BLOODSTONE, or heliotrope, representing the month of March in the list of natal stones, symbolic of courage and wisdom, and the centre of much legendary interest, is one of the most attractive of the green varieties of that almost omnipresent mineral, quartz. The scientific terminology of quartz is involved and complicated by differing authorities in mineralogy, but bloodstone is a massive variety generally classed as plasma, a name, however, that is applied by some to green chalcedony and by others to green jasper; this curious mineral contains spots of red jasper that resemble drops of blood, and to which it owes its name. One of the most striking traditions which concern bloodstone is that it originated at the crucifixion of Christ, from drops of blood drawn by the spear thrust by a Roman soldier into his side, which fell on a piece of dark green jasper. The body of bloodstone is translucent to opaque and of a dark-green colour. Quartz, as is mentioned elsewhere in connection with its gem-stone varieties, crystallises in the hexagonal system; hardness, 7; specific gravity, 2.5 to 2.8 —the purest kinds 2.65. Pure quartz is silica; the varied colours and characters of the many gem-stone varieties are due wholly or partly to contents of iron, alumina, manganese, nickel, and other chromatic constituents. The red spots in bloodstone are simply oxide of iron. The specific name, heliotrope, is favoured by Dana, among other mineralogists. "Heliotrope" is a word derived from two Greek words meaning "sun-turning," and refers to the belief that the stone when immersed in water would change the image of the sun to blood-red. The water was also reputed to boil and upturn the experimental utensils containing this submerged weird mineral.

This opaque, but slightly lustrous, jaspery quartz, although a beautiful and interesting mineral, is not extensively used now in jewelry, and a requisition for it is usually an idiosyncrasy, or because it is a natal stone for those who were born in the month of March. Hardy, tough, yet carved with facility, it is well adapted to signet rings and is usually seen bearing crests or monograms. The ancient Egyptians and Babylonians used the bloodstone extensively for seals. Outside the realm of jewelry it supplies a fine material for artistic cups, small vases, and statuettes. In the French Royal Collection in Paris is a bust of Jesus Christ in bloodstone, so executed that the red spots of the stone most realistically resemble drops of blood. Another fine specimen of carving is a head of Christ in the Field Columbian Museum, Chicago.

The supply of bloodstone is derived almost entirely from India, especially from the Kathiawar Peninsula. Other sources are in Australia and Brazil. Bloodstone does occur, but unimportantly, in Europe; fine specimens are found at several places in Scotland, especially in the basalt of the Isle of Rum.

CHAPTER XXIII

MOSS AGATE

MOSS AGATE is a variety of chalcedonic quartz that has some vogue in the jewelry of to-day, and is one of the most interesting features of gem mineralogy. Enclosed in this stone are what seem to be long hairs and fibres, usually irregularly interwoven, and having the effect of various species of moss. These branching forms, so imitative of one of the most beautiful of plants, are manganese or iron oxide, and not imprisoned vegetation, or prehistoric insects which really were imprisoned in amber, and have been preserved through ages to furnish food for speculation for latter-day naturalists.

The name agate is derived from the river Achates, in Sicily, now called the Drillo, in the Val de Noto. Theophrastus states that this is where ancient agates were found.

Moss agates and Mocha stones are varieties of crypto-crystalline (obscurely crystalline) quartz of fibrous structure, and are slightly softer and

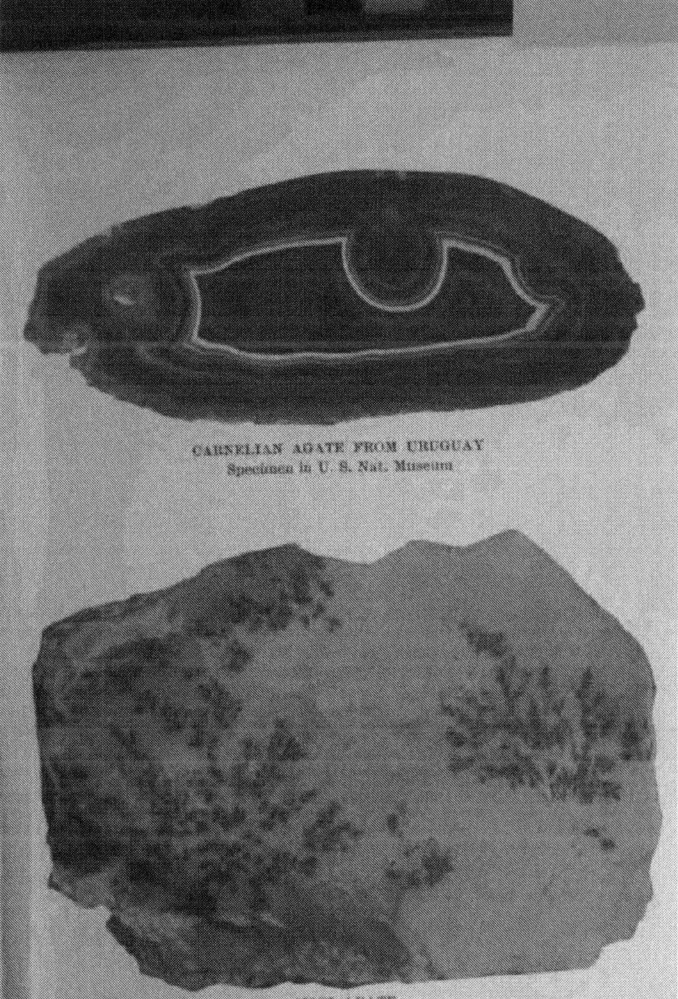

CARNELIAN AGATE FROM URUGUAY
Specimen in U. S. Nat. Museum

MOSS AGATE

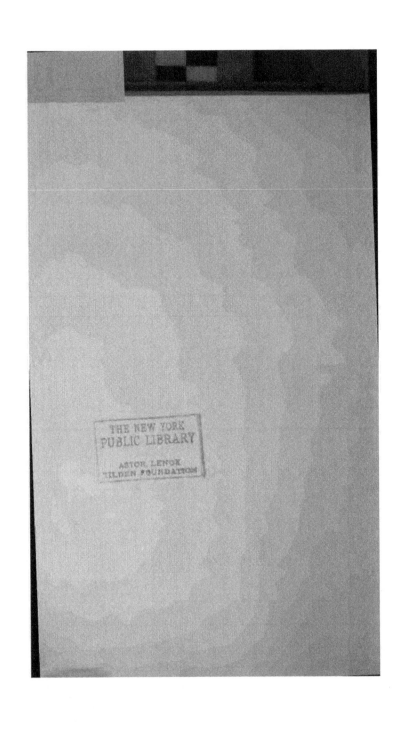

lighter than crystallised quartz. The hardness of crypto-crystalline quartz is 6.5; specific gravity, 2.6; it is more difficult to break than crystalline quartz, being very tough, which makes these varieties—their principal differences being of colour and colour-pattern—eminently fit for carving.

The finest moss agates known to-day come from India, and those specimens called "mocha stone" originally came, it is believed, from the vicinity of Mocha, an Arabian seaport at the entrance of the Red Sea most famous for its aromatic coffee. The Oriental moss agates are common in the volcanic rocks (trap rock) of western India, occurring with Mocha stone. Blocks weighing as high as thirty pounds have been obtained. It occurs also as pebbles in many Indian rivers. From China has come, during recent years, a supply of natural green and artificial yellow and red moss agates, which have, to a considerable extent, replaced others on the market. Fine moss agates are abundant in various parts of the Rocky Mountains; the best are found in the form of rolled pebbles in the beds of streams. As souvenirs, and for sentimental reasons of local interest, these beautiful gem stones of our Rocky Mountain

States are cut and mounted; in the tourist the Western jeweller and curio-dealer finds for these American moss agates a good customer.

Mocha stone ("tree stone" or dendritic agate) is a white or grey chalcedony showing brown, red, or black dendritic markings resembling trees and plants. These have been formed by the percolation of a solution containing iron or manganese through the fine fissures of the stone, and the subsequent deposition of the colouring matter originally held in solution. The brown and red markings are caused by oxide of iron, and the black by oxide of manganese.

Agate in general is but little used in modern jewelry, but for art objects and interior architectural decoration it is always in demand. For centuries, the centre of the industry of cutting and polishing agate has been Oberstein, Germany; an authentic record shows that this industry has existed there since 1497; the industry has for many years been shared by the neighbouring town of Idar. The subject of agate, its origin, mining, treatment, and use in the arts, might worthily supply material for an extensive book.

AGATE WITH CONCENTRIC RINGS

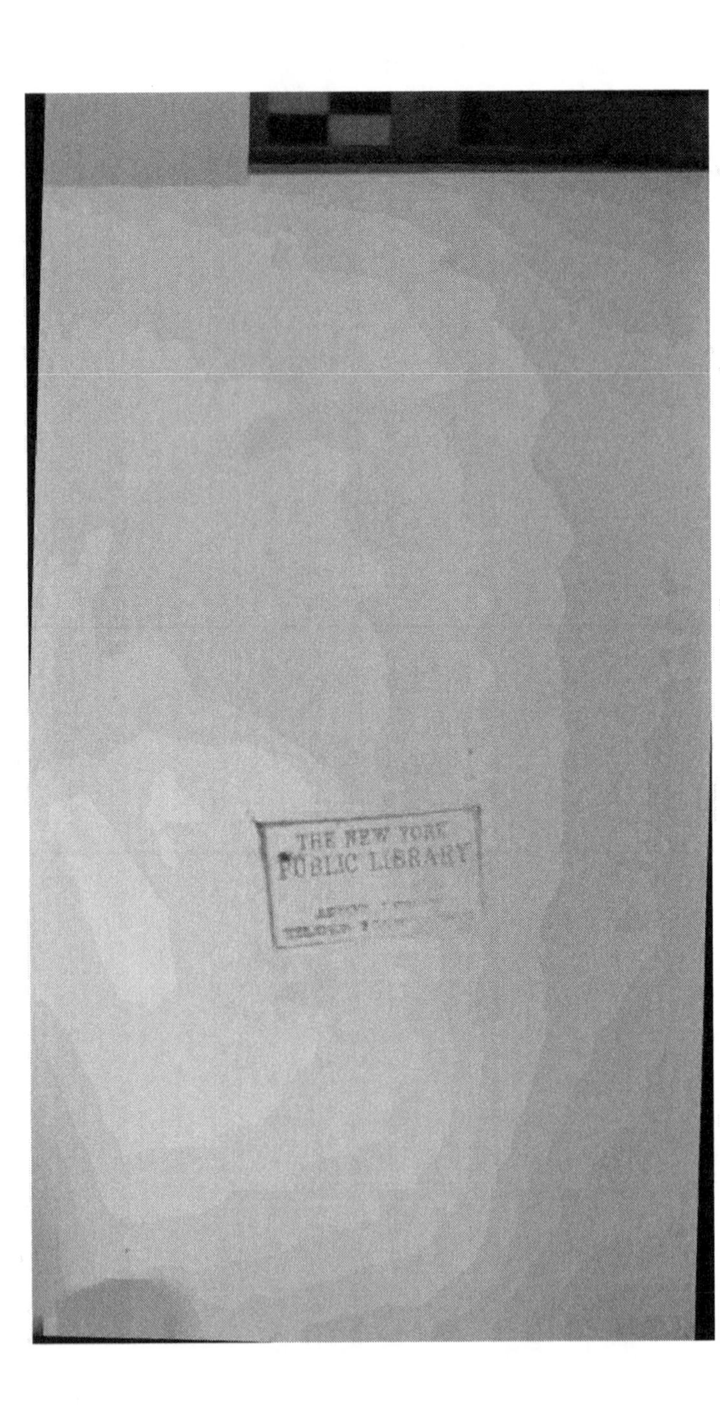

CHAPTER XXIV

ONYX AND SARDONYX

ONYX and sardonyx are varieties of agate with layers in even planes of uniform thickness, thus adapting them to the purposes of cameo engravers. The cameo has a base of one colour and the figure of another. The art of cameo engraving attained a point nearest perfection with the ancient Romans, evidence being supplied by the numerous relics, that are the admiration of modern artists. The word onyx means a finger-nail, and was suggested, it is supposed, by a fancied resemblance to the lustre and appearance of a finger-nail. Of course —if the Greek myth be true—this most beautiful instance of stratification in all mineral nature owes its origin to the freak of playful Cupid, and is the only visible and palpable evidence we have of the mundane visits of the Goddess of Beauty.

Sardonyx is a variety of onyx in which one layer has the brown colour of sard. Chalcedonyx and carnelionyx derive their names from the colours of the intervening layers. "Mexican onyx," it should be noted, is calcite, not quartz, and is very much softer than the real onyx. Mexican onyx has a similar banded structure to real onyx, and is well adapted to architectural or interior decoration, for which it is extensively used, but it is outside the realm of precious stones.

Because of their porous nature, varieties of agate can be easily artificially coloured, and this art has been developed to perfection in Germany, where some of the processes, as "trade secrets," are important phases of the general agate-preparing industry at Oberstein and Idar. The art of colouring agate, which naturally is mostly of a dingy grey colour, was derived from old Rome. Brazilian agate, the material extensively worked now in Germany, is softer than the German varieties that formerly constituted the principal supply, and is particularly susceptible to successful colouring by the scientific German processes.

The onyxes best suited for cameo engraving, besides onyx proper, are chalcedony-onyx, carnelian-onyx, and sardonyx. These are cut so as to display a white or light figure against a darker coloured background. Cameos are mostly engraved in Paris and Italy, but the plates of onyx used by these cameo engravers are prepared at Oberstein and Idar. The tool of the cameo engraver is known as a style.

Perhaps the most famous stone cameo in history was that sardonyx upon which Queen Elizabeth's portrait was cut, set in the famous ring which she gave the Earl of Essex as a pledge of her friendship. When sentenced to death, Essex sent this ring to his cousin, Lady Scroop, to deliver to Elizabeth. By mistake the messenger gave the ring to Lady Scroop's sister, Countess Nottingham, an enemy of the Earl; the vengeful Countess did not deliver the talismanic ring, and in consequence the fated Earl was executed. The Countess Nottingham confessed this act of vengeance to Elizabeth when the Countess was on her death-bed; which, according to the chroniclers of Elizabeth's life history, so infuriated the Queen that she shook the dying noblewoman, saying, "God may forgive you, but I cannot."

Sardonyx—supposed by the ancients to be an entirely different mineral from onyx—was believed to have the power of conferring eloquence upon its wearers; it symbolised conjugal bliss. In *Revelations* it is named as one of the stones in the foundations of the Holy City.

CHAPTER XXV

SEMI-PRECIOUS STONES OCCASIONALLY USED

THE mineral world contains many beautiful materials that are without the pale which encloses the clearly defined gem stones; these "outlanders" may be classed as semi-precious stones that are only occasionally used, and while many are truly beautiful and others are interesting, because of rarity or peculiarities, all lack some quality—usually a sufficient degree of hardness—which would admit them into the patrician rank of Precious Stones. Because of their intense scientific interest, technical mineralogists, who have written books about gems, not only include but devote considerable space to minerals that will not meet the eye of one manufacturing jeweller or gem dealer in one hundred, or ever be seen by one gem buyer in thousands. These stones are usually not so rare in nature as they are in stores, and their cutting and mounting is usually the result of an individual order; otherwise they are col-

AZURITE is a variety of carbonate of copper which shows various shades of azure, merging into Berlin blue. Azurite is both opaque and soft—hardness, 4—and these characteristics limit its use for gem purposes.

BENITOITE. A newly discovered gem mineral of California, blue in colour, and said, when selected crystals are cut in the right direction, to rival the sapphire in colour and to excel the blue corundum gem in brilliancy. The mineral is dichroic, the ordinary ray colourless, the extraordinary ray blue. Benitoite crystallises in the hexagonal system, trigonal division; its most common habit is pyramidal; cleavage, imperfect pyramidal; fracture, conchoidal to subconchoidal; hardness, 6 to 6½; highly refractive. Benitoite fuses to a transparent glass at about 3. It is easily attacked by hydrofluoric acid. Chemically, benitoite is a very acid titano-silicate of barium. Benitoite was discovered in 1907 by Mr. Hawkins and T. Edwin Sanders in the Mt. Diabolo range near the San Benito-Fresno County line. The mineral was determined at the University of California, and is described in a bulletin of its geological department by George Davis Louderback and Walter C. Blasdale.

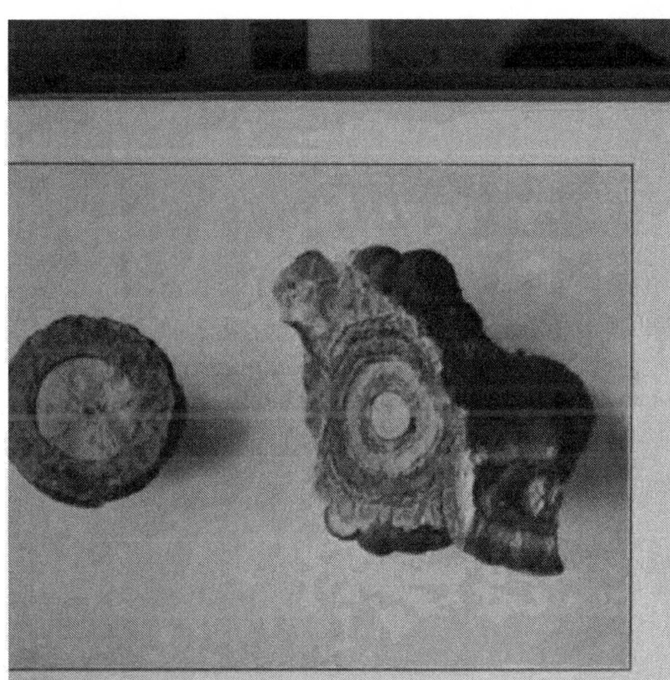

BANDED NODULES OF AZURITE AND MALACHITE
Specimens in U. S. Nat. Museum

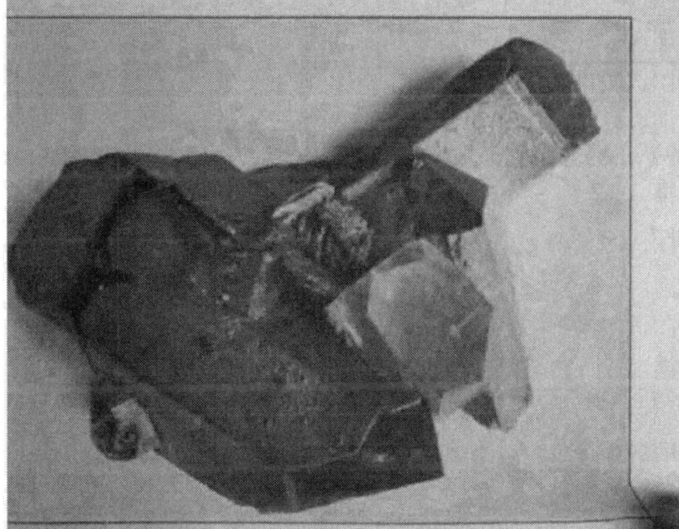

CAIRNGORM is the brown variety of rock crystal, also called "smoky topaz." Cairngorm has a sentimental and historic interest involved in its use as an ornament for the weapons and picturesque clan dress of the Scottish Highlanders.

CARNELIAN is a reddish variety of chalcedony, merging into greyish red, yellow, and brown; it is translucent, like horn. Carnelian takes a high polish and its colours are sometimes heightened by exposure to the sun or by heat. This attractive semi-precious stone was formerly much more extensively used than now, and its merits may, through the vagaries of fancy and fashion, which govern the fates of all gems, again raise it higher in popular favour.

CHONDRONITE, a mineral that is found abundantly at the Tilly Foster mine in Brewster, Putnam County, New York, appears in deep garnet-red crystals of great beauty. Chondronite is classed with the minor gems, and it deserves a more extensive use. Hardness, 6.5. It has a vitreous lustre.

DIOPSIDE is a variety of pyroxene; hardness, 5 to 6; lustre, vitreous or greasy; transparent to translucent; and doubly refractive. Fine specimens, fit for gem purposes, are found near DeKalb, St. Lawrence County, New York. When cut brilliant, diopside makes a very attractive stone and resembles green tourmaline.

DIOPTASE is a silicate of copper; other names for it are achirite and Congo emerald; hardness, 5. The softness and brittleness of this attractive stone disqualify it for extensive use.

FLUORITE or fluorspar, of which chlorophane or cobra stone is a variety, is a highly lustrous, brittle crystal of wide colour range; hardness, 4. Varieties of fluorspar are sometimes termed, in the trade, "false" ruby, emerald, sapphire, and other well-known gem stones.

GOLD-QUARTZ—in crystals, filiform, reticulated, and arborescent shapes—is commonly worn as a jewel. Gold penetrating white, black, rose, and amethystine quartz, is worked into jewelry of all sorts, sometimes of very elaborate designs. These uses of gold-quartz are most common on the Pacific coast and in western North American cities.

HEMATITE, composed of iron 70, oxygen 30, is commonly cut into beads, charms, and intaglios. Chromic iron and ilmenite are similarly used. Although this iron ore is steel-grey,

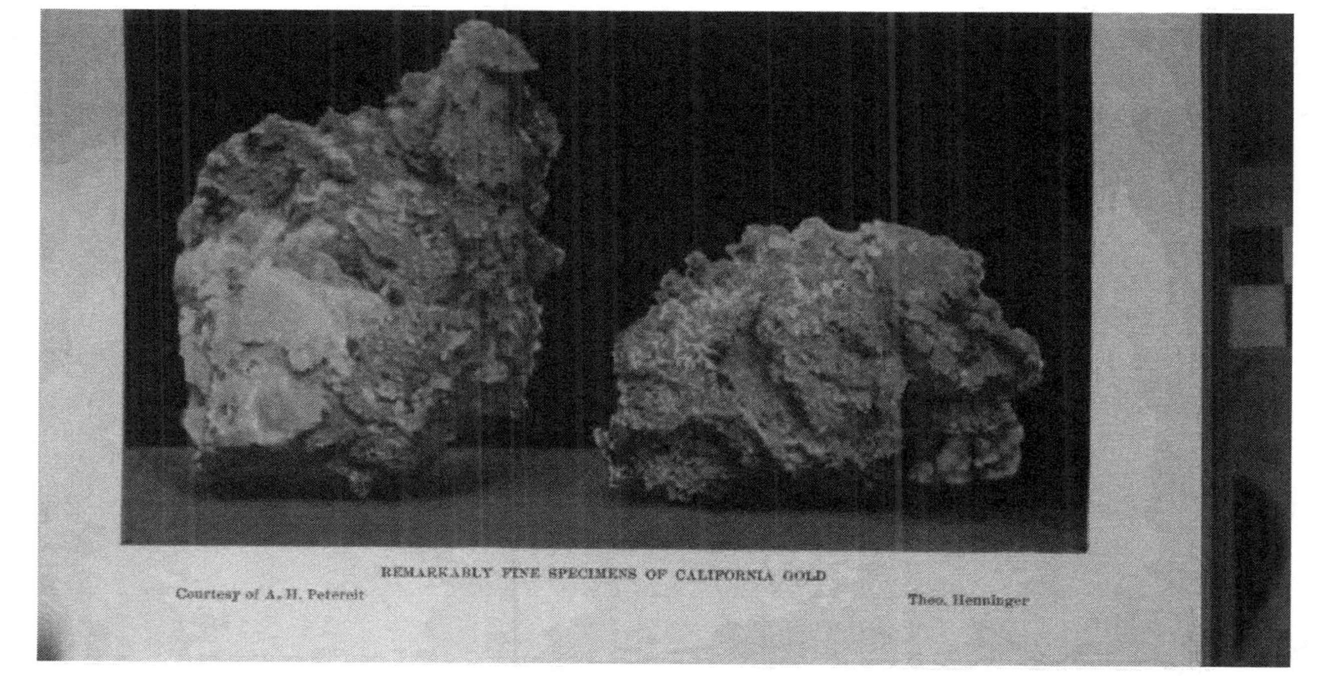

REMARKABLY FINE SPECIMENS OF CALIFORNIA GOLD

Courtesy of A. H. Petereit — Theo. Henninger

when polished, its streak, when scratched, is red; hence the name hematite, meaning "blood-stone."

IOLITE, also called dichroite and water sapphire, is a pleochroic mineral occasionally cut for gem purposes. It is somewhat harder than quartz.

JET is a soft compact light coal of a lustrous velvet black colour, and can be highly polished. It is used not polished for mourning goods. Jet was the agates of the ancients, their source of supply being near the river Gagas in Syria, from which the name of the mineral was derived.

LABRADORITE, sometimes, in the trade, called labrador, is a feldspar. Because of its structure some of the varieties of labradorite reveal a wonderful variety of colours. Labradorite can be highly polished and exhibits beautiful chatoyant reflections.

LAPIS-LAZULI was long regarded as a separate specific mineral; it was the sapphire of the Greeks, Romans, and the Hebrew Scriptures. Instead of being a simple mineral, lapis-lazuli consists of a bluish substance (lazurite) with granular calcite, scapolite, diopside, amphibole mica, pyrite, etc. The hardness of lapis-lazuli is 5.5; specific gravity, about 2 to 4; lustre, vitreous; translucent to opaque.

LAVA can hardly be classed as a semi-precious stone, but it is and has been quite extensively utilised in jewelry, chiefly on account of sentimental association with, and as souvenirs of, volcanoes. Lava is the fusion of various mineral substances due to the heat and force of eruptions from the interior of the earth; it varies in structure and constituents, but the surface lava is usually massive with vesicular or porous marks; fracture, splintery and conchoidal; lustre, dull or glistening; it is opaque and of various colours and shades. Lava frequently contains crystals—feldspar, lenate, hornblende, garnet, and other minerals. Vesuvian lava of a blue tint resembles transparent enamel, and is mounted in brooches and rings; cameos and intaglios are sometimes cut on it.

MAGNETITE, or lodestone, possessing polarity, is used for charms, because of the mystical properties attributed to it.

MALACHITE is carbonate of copper of a bright green colour. When this copper ore occurs in conjunction with azurite, the companion minerals are cut together, with a pleasing effect.

OBSIDIAN is compact volcanic glass, and is

cut for gem purposes to a greater extent than some of the other semi-precious stones here referred to. Varieties are moldavite, or bottle stone, of a green colour; marekanite or mountain mahogany, a red or black and brown banded kind; and Iceland agate; pearlylite; and sphærulite.

PHENACITE, of gem quality, is transparent, colourless, and of a vitreous lustre. This brilliant mineral is harder, heavier, and more refractive than quartz, which it so closely resembles, so that it was not until 1833 that mineralogists differentiated between them. Its name, phenacite, is derived from the Greek word *phenax*, meaning a deceiver. Phenacite remotely resembles the diamond in its brilliancy and refractiveness. Some specimens exhibit pale-rose and wine-yellow colours.

PYRITE is a brass-yellow mineral of metallic lustre known to jewellers as sulphur-stone and technically as marcasite. It is a common mineral, and is so frequently mistaken by the uninformed for gold that it has earned the sobriquet "fool's gold." Pyrite is a sulphide of iron. Although so common as to have no intrinsic value, pyrite constantly remains in use in jewelry and is seen in rings, brooches, and at Matura, Ceylon. Colourless or smoky zircons are called jargons or jargoons. Transparent zircons of a brownish, red-orange colour are called hyacinth or jacinth. Zircon is the heaviest gem mineral—more than four times the weight of water—its specific gravity being [illegible] to 4.86. Its hardness is 7½. So high is its index of refraction—1.92—that it approaches the diamond in brilliancy when cut. Zircons of gem quality come mostly from Ceylon, where they are found in the form of rolled pebbles. Zircon is found in various American localities, but it is opaque.

DIAMOND CUTTER AND SETTER AT WORK

DIAMOND SAWING MACHINES

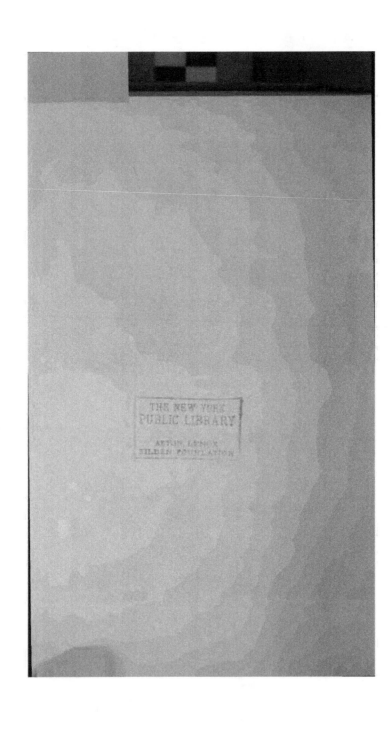

CHAPTER XXVI

CUTTING DIAMONDS AND OTHER GEMS

PRECIOUS stones in the rough are seldom things of beauty. The most valuable gem stones might be dismissed with contemptuous glance by an inexperienced finder, as no doubt has often been the case. Ancient gems that have been benefited only to the extent of the crude handiwork of the artisans of their period, reveal but little of the imprisoned chromatic beauty and flaming splendour that would make them magnificent under the scientific and artistic treatment of a modern diamond-cutter or lapidary. Thus the work of the highly skilled [illegible], who cut diamonds, with their co-[illegible], who set the diamond in a tool with which the cutter applies the rough stone to the [illegible] wheel, and the toil of the lapidary, [illegible], forms, and polishes semi-precious [illegible] of the greatest importance in making [illegible] the beauty and value of gems. Here [illegible] that the craft of the diamond cutter and the trade of the lapidary are absolutely separate and distinct in the methods that each employs in cutting and polishing gem minerals. The diamond cutter cuts diamonds only. The lapidary cuts and polishes all other precious and semi-precious stones. Both diamond cutter and lapidary prepare the way for the craft of the jeweller, to whose judgment and art in design and manufacture the cut gem owes its environment, which will go far to increase or mar its beauty. For the jewellers' art is as important to the gem as the scenic artist's and stage manager's is to the actor's dramatic art; and without intelligent co-operation, the jeweller might detract from the appearance of a gem that the capable diamond cutter or lapidary has done so much to enhance.

Thus the cutting of gem stones is necessary for the full development of the inherent properties upon which their beauty is dependent. A gem, as extracted from the earth, may be opaque, irregular in form, and contain superficial flaws and imperfections; but when relieved of its incrustations and reduced to a size that would permit of the elimination of its imperfect portions, it becomes transparent and its imprisoned fires are released in brilliant flashes.

Occasionally a gem does appear which, without artifice, may plainly show its qualifications for high rank in the court of gems; but, in the main, the development of its beauty to a high degree necessitates cutting and polishing. The highly specialised work of the diamond cutter or lapidary involves compliance with geometrical principles and rules; adaptation to the place occupied by the gem stone under treatment; a knowledge of the clearly defined science of crystallography, especially with regard to the planes of cleavage; careful consideration of the stone's degree of hardness, brittleness, and a thorough acquaintance with the established forms of cutting and the results achieved through them with different kinds of gem minerals and their chromatic varieties.

The art of gem-cutting has progressed gradually from the crudest beginning. Man's first attempts to artificially improve the appearance of gem stones extended only to polishing the natural surfaces; later, the worker essayed to round the rough corners, and in the course of the evolution of this art, efforts were made to reduce the stone to a symmetrical shape. Gem-cutting by Oriental workmen, in the island of

Ceylon, Burma, and India, has, even now, advanced but little beyond its crude beginnings. The Asiatic artisan uses a polishing disc on the left end of a horizontal wooden axle, which revolves in sockets on two upright pegs driven into the earth or set in the timbers or boards which floor his dwelling or shop. The motor for this machine is a long stick to which a cord is tied, as to a bow, at each end, one turn having been taken around the axle; the motive power is supplied by the right hand and arm of the operator, who moves the stick back and forth; there is usually no holding tool; the stone is held in the fingers of the left hand and thus pressed against the surface of the polishing wheel. The abrasive powders of corundum or some mineral nearly as hard, mixed with water to a paste of suitable consistency, are at hand, contained in the halves of cocoanut shells. The earliest record of the artificial improvement of gems by the ancient Greek and Roman artisans proves them to have had higher ideals and more invention than Orientals, especially in the matter of imparting to stones symmetrical forms; the greatest advance they made, however, in the treatment of gem minerals, was in their art in cutting cameos and intaglios, their engraving

ORIENTAL GEM CUTTERS

of gems having early reached a surprisingly high state of perfection.

The centres of the art and industry of diamond-cutting are at Amsterdam in Holland and Antwerp in Belgium, but the very highest form of the art was initiated in and is practised in these United States; here, without senseless waste and extravagance, the intrinsic value of precious stones, as determined by their weights, is sacrificed to artistic effect, beauty, and brilliancy. This high degree of gem treatment is in strong contrast with the more economical practice in Europe, and is the antithesis of the custom in Oriental countries, where weight is conserved at the expense of brilliancy and beauty.

The styles of cut may be grouped as follows: 1, those bounded by plane surfaces only; 2, those bounded by curved surfaces only; 3, those bounded by both curved and plane surfaces. The styles of the first group are best applicable to transparent stones, as the diamond, emerald, and ruby; they are brilliant cut, double brilliant or Lisbon cut, half brilliant or single cut, trap or split brilliant cut, Portuguese cut, star cut, rose cut, or briolette, step brilliant or mixed cut, table cut, and the twentieth-century cut; this is a combination of facets that was experimented with but not very successfully about the year 1903. Styles of the second and third groups are best adapted to translucent and opaque stones, such as the opal, turquoise, moonstone, and cat's-eye. Both the first and second styles are applied to garnets, which are cut either with facets or convex (or *en cabochon*), and when thus cut they are termed carbuncles. The styles of the second group are bounded by curved surfaces; they are the single cabochon cut, double cabochon cut, hollow cabochon cut, and tallow top cabochon cut. The third division of styles are those bounded by curved and plane surfaces, represented by the mixed cabochon cut.

The brilliant cut could be represented by two truncated pyramids, placed base to base; the upper pyramid, the *crown*, is truncated in a manner to give a large plane surface; the lower one, the *pavilion*, ends almost in a point. The line of junction of the bases of the two pyramids is called the girdle. While there are many modifications of this style, as to the size, mutual proportions, and number of facets, the facets in the perfect brilliant number fifty-eight. The top facet is called the *table*, and is formed by removing one third of the thick-

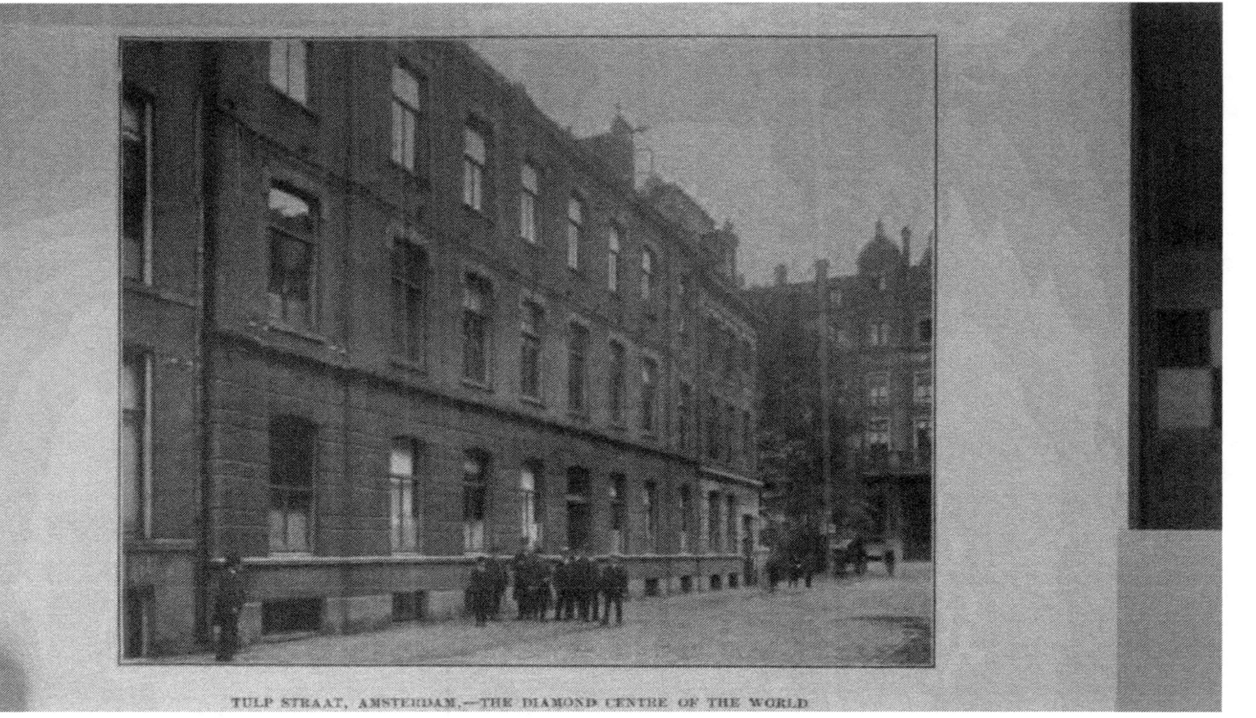

TULP STRAAT, AMSTERDAM,—THE DIAMOND CENTRE OF THE WORLD

ness of the fundamental octahedron; the bottom facet is called the *culet,* or *collet,* and is formed by removing one eighteenth part of the stone's thickness. The triangular facets touching the *table* or summit of the crown are called *star* facets; those touching the girdle are divided into two groups, *skill* facets and *skew* facets. The corner facets touching the table and the girdle, when on the crown, and the culet and girdle, when on the pavilion, are called, respectively, *bezel* or *bizel* facets, and *pavilion* facets. A summary of the number of facets and their distribution is as follows: 1 table, 16 skill facets, 16 skew facets, 8 star facets, 8 quoins, 4 bezel facets, 4 pavilion facets, and one culet. Sometimes the cut is modified by adding extra facets around the culet, making sixty-six in all.

The brilliant cut is especially applicable to the diamond; when perfect it should be proportioned as follows: From the table to the girdle, one third, and from the girdle to the culet two thirds of the total. The diameter of the table should be four ninths of the breadth of the stone. These proportions when applied to other stones than the diamond are modified to suit the individual optical constants of the gem.

The double brilliant, or Lisbon cut, is a form with two rows of lozenge-shaped facets, and three rows of triangular-shaped facets, seventy-four in all.

The half brilliant, single, or old English cut is the simplest form of the brilliant and is now generally employed for small stones; when the top is cut so as to form an eight-pointed star it is called the English single cut.

The trap brilliant, or split brilliant, differs from the brilliant in having the foundation squares divided horizontally into two triangular facets, forty-two in all.

The Portuguese cut has two rows of rhomboidal and three rows of triangular facets above and below the girdle.

In the star cut the table is hexagonal in shape, and is one fourth of the diameter of the stone; from the table spring six equilateral triangles, whose apexes touch the girdle, and these triangles, by the prolongation of their points, form a star.

The crown of the rose cut consists of triangular or star facets, whose apexes meet at the point or crown of the rose. The base lines of these star facets form the base lines for a row of skill facets whose apexes touch the girdle, leaving spaces which are cut into two facets. The base may be either flat or the bottom may be cut like the crown, making a double rose or briolette cut. The shape of a rose-cut stone may be circular, oval, or, indeed, any other to which the rough stone may be adapted.

In the trap or step cut, the facets extend longitudinally around the stone from the table to the girdle, and from the girdle to the culet. There are usually but two or three tiers of step facets from the table to the girdle, while the number of steps from the girdle to the culet depends upon the thickness and colour of the stone. This style of cut is best adapted to coloured stones.

The form of the step brilliant, or mixed cut, from culet to girdle is the same as that of the trap cut, while from the girdle to the table the stone is brilliant cut, or the opposite.

The table cut consists of a greatly developed table and culet meeting the girdle with bevelled edges. Occasionally the eight-edge facets are replaced by a border of sixteen or more facets.

The twentieth-century cut contains more facets than the brilliant and is differently shaped and arranged. Originally this style was designed with eighty-eight facets and proportions similar to the American brilliant, but with a greater height from the girdle to the centre of the table, caused by the facets replacing the table being carried to a low pyramidal point in the centre. Subsequently the style was modified, the stone being cut thinner and with but eighty facets, the central top facets being almost flat. This cut is helpful in some cases, especially to shallow stones, but it probably exceeds the limit of efficiency in the effort to increase the surface reflection and dispersion of light rays, and experience has not demonstrated its success.

The cabochon cuts represent different degrees of convexity above the girdle, and beneath a concave, plane, or slightly convex surface. The double cabochon is customarily cut with a smaller curvature on the base than on the crown. The single cabochon is a characteristic cut for the turquoise. The hollow cabochon is best for deep-coloured transparent stones. The mixed cabochon has either the edge or side, or both, faceted. The degree of convexity in the various cabochon cuts is made to depend upon the nature of the stone to which the cut is to be applied. The cabochon cuts are specifically within the province of the lapidary.

The process of cutting gems is simple, but the results are due to the skill and especially to the judgment of the cutter. That part of the surface of a rough stone at which it is desired to place a facet is rubbed with a harder stone or with some other effective substance. The harder stone or substance abrades small fragments and powder from the softer, and gradually the surface of the subject mineral is transformed into a plane face, or facet. In like manner other facets are added or a rounded surface is produced by similar means. In grinding, the harder stone or abrasive material is reduced to a fine powder and mixed with olive oil into a paste (if diamond powder), or with water (if emery), and placed near the edge of a circular disk, or "lap," which is about twelve inches in diameter and an inch in thickness. The lap, usually of metal, revolves horizontally with great velocity, and the precious stone to be ground is pressed against the disk where the disk is loaded with the abrasive paste; the pressure causes the powder to become embedded in the soft metal of the disk. This acts as a file, equal in hardness to the grinding powder. The duration of the operation depends upon the hardness of the precious stone and of the abrasive material. The skill required of the operator involves the most careful watchfulness against exceeding the size prescribed in the plan for the stone; also against overheating the stone, which causes the development of small cracks in the interior of the stone called "icy flakes." An essential prerequisite for grinding precious stones is a means by which they can be held steadily and true in a desired position. For this the diamond-polisher uses a time-honoured tool called a "dop" (commonly pronounced "dub"). This holder of the rough diamond is a small hemispherical cup of iron attached by the convex side to a stout copper rod. The cup is filled with an easily fusible alloy of tin and lead, which is fused and allowed to cool; just before this composition solidifies the stone to be cut is set in the position desired in the cooling alloy, with about half its bulk projecting from the metal. Thus the stone is firmly fixed in an immovable position. The semi-precious stones, when cut by the lapidary, are set in the end of a wooden holder, or "stick," with some kind of resinous cement.

Diamond cutters formerly cut the diamonds in a small wooden box especially designed for this use; all of the operator's strength was needed to rub two diamonds together, a process called "bruting," so that the attrition under this pressure would cut the stone into the shape desired. About the year 1888 the first machine was invented to shape diamonds, and the cutter, who formerly had to cut the stone twice, or several times, accomplishes the same result in one operation. All diamond-cutting in America is now done by machine, while in Europe the smaller sizes are still cut by hand in the old tedious and laborious method. The tools for polishing remained unimproved from the inception of the modern diamond-cutting industry until the year 1896, when the machine dop or holder was invented. This modern machine dop, although still an imperfect device, holds the stone without the application of the mixture of lead and tin, but it can only be used for stones of a fair size. The majority of the cutters and polishers of diamonds in the United States now use these mechanical dops, as the market and industry in America demands stones of considerable size almost entirely; it is impossible to use these dops for the stones of small size exclusively cut in Europe. The inventor of the machine dop also invented the machine for sawing diamonds. Through the use of this device pieces of the stones which were formerly polished away and ground to worthless black dust are now saved. The economy effected by the sawing machine is illustrated by its use in cutting off the apexes of the rough diamond crystals; the smaller parts, called *melée,* are sent back to Europe to be cut.

CHAPTER XXVII

IMITATIONS, IMPROVEMENTS, AND RECONSTRUCTION

COUNTERFEITING precious stones of the higher classes has the same motive as counterfeiting coin or paper money, and is easier, because gems have no official characteristics, the physical and chemical characteristics are known to but few, and the counterfeiter does not hazard the penalties that the stringent laws of all nations enact against counterfeiters of the currency, the deterrent and punitive effects of which, however, despite their severity, have never entirely prevented successful counterfeiting. The counterfeiter of precious stones, and the dealer who knowingly and deceptively sells his product for an undue profit, swindle, and they are amenable to the criminal and civil laws, if evidence can be secured upon which to base successful prosecution and suits, a difficult matter generally, especially to prove guilty knowledge and intent. An enormous quantity of imitation gems is constantly being manufactured and sold under various qualifying terms that preclude the possibility of the purchaser establishing a claim that deception was practised, and in most cases the price paid was far from that which a genuine stone of equal weight would bring in any market. These imitations frequently bring to their buyers one disappointment, in that their brilliancy soon deteriorates or fades almost entirely. Sometimes "diamonds," which are qualified with such prefixes as "Alaska," "Sumatra," "Borneo," or any other name dictated by the dealer's fancy and which, it is hoped, will sound to the ear of a possible customer like a locality where diamond mines might be, are quartz or some other simple mineral; but in general they are of glass that has long borne the time-honoured name of "paste." Merchandise of this peculiar kind is so favourably exhibited in show windows and showcases by electric lights and other advantages, as to deceive the inexperienced prospective buyer. By the merchants who offer for sale these transparent imitations they are called "white stones."

Every gem for which there is a considerable demand has been, is, and, probably, always will be, imitated. Another name for "paste" is "strass," derived from a man named Strass of Strassburg, capital of the province of Alsace-Lorraine, Germany, who invented one of the several formulæ and processes employed to create the brilliant, heavily lead-impregnated glass so enormously used in the counterfeiting of gems. While the many prescriptions for the strass composition vary in constituents and proportions, a fair sample of these mixtures is as follows:

Pure powdered quartz	38.2
Red lead	53.3
Potassium carbonate	7.8

The ingredients are pulverised, mixed, and heated in a crucible with a temperature raised gradually until the compound fuses, with great care. It is maintained at that point for about thirty hours and then slowly decreased. The factors in securing a result that will fulfil all requirements are the thoroughness of the previous mixing, the regularity of the temperature, the duration of the fusion, and the slowness of cooling. The clear paste is cut for imitation diamonds, while for the coloured gems the hue desired is imparted by the solution of metallic oxides and other substances; manganese oxide

Both the genuine and artificial ruby are unaffected, while all imitations made of paste, as imitation ruby, sapphire, emerald, etc., are quickly attacked.

To M. Antony Jacques, a jeweller of Grenoble, France, is accredited the discovery of a new method of detecting counterfeit emeralds and garnets, a method that is simple and that can be applied by any person. Through two coloured glasses, placed across and upon one another, one blue and the other yellow, the stone in question is examined, the stone being placed directly against an electric lamp. The genuine emerald will appear to be of a violet colour, no matter whether it is a "scientific," a "reconstructed" gem, or an ordinary green doublet. The most convincing imitation will appear unchanged and the deception thus easily demonstrated. A genuine garnet similarly placed upon an electric lamp and looked at through pale-green glass will appear decolourised, while a counterfeit will remain a garnet colour. The author's experiments have demonstrated the efficiency and reliability of these tests.

Besides the complete imitation of gems there are partial sophistications in which considerable ingenuity and constructive ability are displayed by creating "doublets" and "triplets." The doublet is constructed with the table and crown of a genuine stone, usually off-coloured, cemented to a pavilion made of a paste having the approved colour, thus giving the valueless crown the appearance of a fine stone. The softness of its pavilion usually betrays the doublet. As a guard against this discovery the triplet was invented. This is a real gem, usually pale or off-coloured, with a thin layer of coloured glass at the girdle. The detection of this combination usually requires the magnifying glass and specific gravity tests; the glass usually betrays the deception, and if soaked in alcohol, carbon bisulphide, or ether, the fraud usually separates. Pearls are imitated by coating the inner surfaces of glass beads with a preparation made from fish-scales.

Substitution of other minerals for specific precious stones has not the shadow of justification that sometimes softens the annoyance of receiving, or being offered, "something just as good" in drugs, groceries, or dry goods. The substitutes sometimes offered or proposed for diamonds include white sapphires, zircon, quartz, and white topaz. Artifice is frequently employed to heighten or change the colour of a real gem by thermal or chemical treatment; thus heat may remove the colour or increase the brilliancy of topaz, sapphire, and other precious stones. Heat will change the colour of a wine-yellow Brazilian topaz to a rose-pink; the same influence may whiten and render more brilliant an off-coloured or spotted diamond. A high temperature will often alter and improve the colour of the cairngorm, citrine quartz, and other minor gems. Chemical solutions can be successfully applied to turquoise to deepen its colour and invest it with permanency; agates are commonly dyed, and by chemical aid colourless chalcedony is converted into an excellent imitation of the moss agate. An off-coloured diamond may be apparently changed to a stone of good water by a wash of aniline blue, but the effect is but temporary. Besides these, the interiors of settings may be backed, stained, or enamelled, usually entirely legitimate improvements.

Far different from the imitation of gems is the making of them by artificial means, with the result of a real gem that is but slightly distinguished from those produced in Nature's laboratory. Although there are distinctions discernible to the expert with the aid of the magnifying glass, the gem stones thus produced —that are worthy of notice—contain the same component parts in their proportions that the natural stones do, and equal them in the principal characteristics of hardness, specific gravity, and refractiveness.

To quote Wirt Tassin:

A sharp distinction is to be drawn between the imitation of a gem stone and its formation by artificial methods. The imitation gem only simulates the natural substance; the artificial gem is identical with it in all its chemical and physical properties. Until recently the laboratory gem was hardly more than a curiosity, although its synthesis has undoubtedly been of value from the theoretical standpoint. Examples of this class are to be found in the diamond as produced by Moissan in the electric furnace and the synthesis of spinel and chrysoberyl by Ebelmen from mixtures of alumina and glucina, respectively, using boric acid at very high temperature as a solvent. Hydrofluoric acid and silicon fluoride have also been used to induce combination between silica and other oxides. In this manner topaz, a complex fluo-silicate, has been made by the action of fluoride of silicon upon alumina.

The minerals thus formed have usually been very small and of no commercial value. Quite recently, however, rubies have been produced by the fusion of alumina with a trace of chromium oxide in the electric furnace, and the art has progressed to such an extent that the product is now on the market for sale as watch jewels. The electric furnace has also produced another product which, while strictly speaking not a synthetic gem, yet is essentially an artificial one. Imperfect rubies, chips, and small stones, are fused in the furnace together with the addition of a small amount of colouring oxide such as chromium. The fused product is then cut and polished, and the result is a ruby of good colour and of fairly large size. Emeralds and other coloured stones have been made in the same way, and so promising has the industry become that the courts have been called upon to decide what constitutes a ruby. Their decision was in substance that the word ruby could be legally applied only to the red-coloured corundum, anhydrous oxide of aluminum, occurring ready formed in nature.

Reconstructed rubies however are in the main rightly placed and justly valued, for they are generally used in large quantities for medium-priced jewelry.

The French chemists Frémy and Verneuil have succeeded in manufacturing true gems, rubies chiefly, but also sapphires, by artificial processes. A title given to gems created by this or similar processes by man is "scientific" ruby, emerald, sapphire, or whatever the gem may be. Mr. Rudolph Oblatt of New York is an American producer of the "reconstructed" ruby, which has attained some commercial success, and its effect upon the market for rubies, whether this be considered desirable or otherwise, has been to lower the price of natural rubies because the demand has been lessened for them; this applying probably only to stones of one carat or less. When "reconstructed rubies" were first offered to the jewelry trade in Paris, and subsequently in the United States, their makers encountered many disheartening rebuffs; to-day many merchants and manufacturers who at first were horrified by, and who resented the suggestion of using the "reconstructed ruby," are complacently handling them in a continually increasing market for medium grade jewelry.

Mr. Oblatt describes his process as follows:

From the small genuine particles of ruby or "ruby sand" found with the real rubies in Burma I select pieces that are alike in colour and qualities; one of these chips I place upon the top of a "U"-shaped platinum iridium tube. Upon this is focussed the heat from two jets of oxygen and hydrogen gas—for the latter can usually be substituted gas from the street mains, as it contains a sufficient proportion of hydrogen gas to qualify it for this use—with a pressure of eight hundred pounds to the inch, producing a temperature of six thousand degrees F. As soon as the first chip is melted I introduce into the flame at the end of an

iridium holder a second chip, which when it melts flies off and adheres to the first melted chip and they are fused together. The continuation of this process of adding particles results in the production of a genuine ruby of the shape of a pear, resting on its stem—the first chips fused—varying from five to ten carats in weight. The operation lasts from one to two hours, according to the size of the stone produced. The most difficult part of the process is the cooling; Nature's laboratory in which the ruby was produced had the resources of a tremendous sustained heat and a cooling process of unknown duration. In general, Nature's cooling process was too rapid, the evidence being in the minute cracks, called ribbons, which run through most rubies and the absence of which makes the perfect ruby one of the rarest and costliest of stones, especially when the cut gem weighs two carats or more. The cooling process is secret and one of the most important factors in the achievement of the reconstructed ruby. The enlarged ruby is then cut by the lapidary exactly as is the natural ruby, for it is the same in its chemical and physical constitution. This is attested by analysis made by very high scientific authorities, their reports being in my possession and open to the inspection of anyone.

The scientific ruby is wholly the result of artificial means but is genuine to the extent of being a properly proportioned combination of the chemical constituents of the natural ruby; in manufacturing the scientific ruby we begin with a solution of common alum, to which a trace of chrome alum is added as the ultimate colouring constituent. Now add ammonia, and there results a gelatinous precipitate of the hydrates of aluminum and chromium. This gelatinous precipitate is filtered off, evaporated down to dryness, and subsequently calcined into an intimate mixture of alumina and the oxide of chromium. It is then ground into an impalpable powder, and placed in the transforming apparatus. Through a tube passes a supply of coal-gas, through another tube a supply of oxygen. The two meet where they are ignited, and constitute a carefully regulated flame whose temperature is practically two thousand degrees. In a box at the top, is placed the powdered alumina, and the bottom of this box consists of a fine sieve. A small automatic tapper carefully jars the powder through the sieve and through a tube, which serves for the supply of oxygen. It thus happens that every trace of the powder must pass through the flames of two thousand degrees.

In a critical review of this process and its results, a very high scientific authority stated that:

These properties agree exactly with those of the natural ruby; but there is one feature by which these stones could be recognised as having been artificially produced; and that is by the form of the cavities existing in them, these being always spherical. The cavities in a natural ruby are always of an irregular form, and this would always afford a means of detecting the artificial stone.

The stones are rubies and are not imitations, as so many of their predecessors have been. But they are not natural rubies, even although produced from clippings of the same, since the crystalline growth is a new one after the clippings have been fused.

The sapphire as well as its sister of the corundum family, the ruby, has for years been the object of solicitude on the part of scientific experimentalists, who would produce real sapphires by artificial means; Mr. A. H. Petereit, of New York City, the well-known dealer in gems and gem minerals, who purveys rarities in this line to collectors the world over, and whose inventive genius is represented by more than twenty-five patents, exhibited to the author a "reconstructed sapphire" which, tested merely by a visual examination, rivalled natural sapphires, that of the same colour and purity would be very costly gems. Mr. Petereit's process is secret, and he modestly claims success only to the degree of producing stones of a size that will cut into small gems. Of the Petereit sapphires *The Mineral Collector* says:

We are pleased to announce that the honour has fallen to an American to at last manufacture a real reconstructed sapphire; successful in hardness, colour, brilliancy, and transparency. Efforts have been made in France, Germany, and other countries to successfully make blue sapphires, and, although they have been successful up to the cooling point, they always lost their colour and became gray when cool.

Mr. A. H. Petereit has had a German chemist working on a formula of his own for two years past, and has had his efforts at last crowned with success. At a meeting of experts in the gem business the reconstructed sapphires were placed among the real stones and they had to admit they were equal if not superior to the real gems.

When Mr. Petereit took up the mineral business his inventive mind was turned into a new channel, the manufacture of artificial gems. Already stories were being told of great successes accomplished in this line, but when it came to produce the stones they failed in one form or another; either the colour or hardness was wanting.

The new sapphires he has invented are perfect in every way. The cannot be scratched by the natural sapphire, they have a beautiful deep blue colour, their brilliancy is only equalled by the diamond, their specific gravity is exactly the same as the natural stone.

His success with scientific rubies was due to the fact that those he handled were the best in the market. They were made from small natural stones by a secret process and not from aluminum and other chemicals, as many now on the market were.

The *Deutsche Goldschmiede Zeitung*, a German jewelry trade journal, has supplemented an article, from which we quote, published upon the points of difference between reconstructed and genuine rubies, by presenting some additional facts, and especially by reproducing two illustrations made from enlarged photographs of reconstructed and genuine rubies supplied by A. F. Kotler, of St. Petersburg:

On careful examination, in the case of the artificial ruby, we notice at once the typical concentric lines as well as the little bubbles occurring in large numbers, which are always spherical, having, in other words, the character of an air bubble in a melted mass. The concentric fine lines, showing variations in the colour, were compared at the time with the circular or spiral lines that result from the string of a paste-like mass, leaving nothing to be desired as far as plainness is concerned. A naturally formed genuine ruby also shows spaces or enclosures, but these are more or less angular, being bounded by crystalline surfaces, The angularity of these voids is, moreover, determined by the entire crystalline structure of the natural stone.

If, therefore, in the genuine ruby, the colour is unequally distributed, the colour stripes *invariably* assume a *vertical* direction, are *never concentric* as in the artificial stone. We may also frequently note that the colour does not run in one direction, but that colour stripes, often of varying intensity, cross one another at obtuse angles; in other words, correspond strictly with the crystalline structure of the grown stone. We may reiterate the assertion that in a genuine natural ruby concentric lines are *never* noted. This most important, and at the same time certain and simplest, distinguishing characteristic, is the more to be regarded, inasmuch as the specific gravity, the colour, the hardness, and the dichroism—in other words, all the optical and chemical properties—of the artificial ruby correspond, more or less, with those of the genuine stone and consequently the scientific assistance, in this case, fails us entirely. An experienced gem expert will, moreover, recognise the genuine ruby by its peculiar, characteristic, soft, silky brilliance, which is lacking in all artificial rubies.

At the recent convention of German jewellers in Heidelberg, where the question as to the nature of the so-called artificial or "scientific" precious stones was exhaustively discussed and a resolution expressing an attitude of opposition towards excessive advertisement of these productions was adopted, Court Jeweller Th. Heiden, in the name of the "Association of Jewellers, Gold and Silversmiths of Bavaria," spoke in favour of hearing an opinion of a prominent authority in regard to the entire subject. According to the *Journal der Goldschmiedekunst*, this has now been rendered, the well-known mineralogist Prof. Dr. Conrad Oebbeke, of the technical high school in Munich, having expressed himself as follows, concerning artificial precious stones:

Between the natural and the artificial precious stones, the material difference will always exist, that one is a natural, the other an artificial product. Up to the present time, I have not seen a *single artificial precious stone that could not be recognised as such.* The claim that the artificial stones are not to be distinguished from the natural gems, that they are absolutely free from defects, etc., according to my experience, is *not justifiable.* Even if it is possible to produce precious stones having the same crystallographic, physical, and chemical properties as the natural gems, they are nevertheless *not equal in value* to the natural product. No more so than an ever so carefully executed and deceptively similar copy of a work of art, a painting, a piece of sculpture, etc., can be called the original. The artificial products, made in the laboratory, are not formed under the same conditions as the natural article, and for this reason we may rest assured that, even should the present scientific methods of distinguishing the genuine from the artificial precious stones fail, further scientific investigation will reveal a method that will make the distinction possible. Interesting as may be the success thus far attained in the production of artificial precious stones, and while we may congratulate ourselves on the progress made in chemical technics in this direction, to the *connoisseur, these articles will always be artificial products* that can never deprive the natural stones

of their value. On the contrary really beautiful natural precious stones will *only be the gainer.* The claim that synthetic stones will ever break the market for real precious stones, is, in my opinion, utterly unfounded.

CHAPTER XXVIII

FOLK-LORE

BECAUSE of their density and hardness, gems are among the most permanent of substances, and yet, to a greater degree, perhaps, than any other kind of property, their value rests on sentiment. The associations of gems in the human mind are so numerous and varied, that no writer has ever attempted to assemble all of them; some are well substantiated in history, others only in legend; they are identified with many religions, but most of them are black with superstition, its origin generally obscure. This phase of the general subject of gems can be properly covered under the term and title of "folk-lore." The Bible's many references to gems are familiar alike to Hebrews and to all Christian readers of Holy Writ. Besides the scattered references and metaphorical use of the names of gems, the Bible contains three lists of precious stones. The first is an account of the jewels on the *ephod*, or short two-piece coat of Aaron, the Jewish High Priest, to the front of which was attached the sacerdotal breastplate. The front and back parts of this coat were united at each shoulder with an onyx mounted in gold and engraved with the names of the tribes of Israel, six on each stone, in memory of the promise made by the Lord to them. (Exodus xxviii., 6, 12, 29.) The breastplate was made of the same material as the *ephod*, and folded so as to form a kind of pouch in which the *Urim* and *Thummim* (Light and Perfection—according to one version) were placed. (Exodus xxxix., 9.) The external part of this gorget, or "breastplate of judgment," was set with four rows of gems, three in each row, each stone set in a golden socket and having engraved upon it the name of one of the twelve tribes of Israel. (Exodus xxviii., 17–20.)

The names of these stones, taken from Biblical antiquities by Adler and Casanowicz, and written for the *Report of the United States National Museum*, for 1896, page 943, are given as in the original and in the Septuagint, together with the meaning agreed upon by most authorities. The rendering of the Revised Version, both in text and margins, is added in parentheses, the list being as follows: 1. *Odem* (*sardion*), carnelian (sardius, ruby). 2. *Pitdah* (*topazion*), topaz or peridot. 3. *Bareketh* (*smaragdos*), smaragd or emerald (carbuncle emerald). 4. *Nofek* (*anthrax*), carbuncle, probably the Indian ruby (ruby, carbuncle). 5. *Sappir* (*sapfeiros*), sapphire or lapis lazuli (sapphire). 6. *Yahalom* (*jaspis*), onyx, a kind of chalcedon (diamond, sardonyx). 7. *Leshem* (*ligyrion*), jacinth, others sapphire (jacinth, amber). 8. *Shebo* (*achates*) agate. 9. *Achlama* (*amethystos*), amethyst. 10. *Tarshish* (*chrysolithos*), chrysolite, others topaz (beryl, chalcedony). 11. *Shoham* (*beryllion*), beryl (onyx, beryl). 12. *Yashpeh* (*onychion*), jasper.

It should always be borne in mind that in many instances the equivalent of the Biblical names of gems is uncertain in the nomenclature of modern mineralogy, therefore there are several lists of names given for the stones in the breastplate. There is an ancient silver breastplate employed as an ornament for the MS. copy of the Torah, or Pentateuch, used in an ancient synagogue, preserved in the Division of Oriental Religions in the United States National Museum. According to this exhibit the twelve stones, with the names of the twelve tribes, are as follows: Garnet, Levi; diamond, Zebulon;

amethyst, Gad; jasper, Benjamin; chrysolite, Simeon; sapphire, Issachar; agate, Naphthali; onyx, Joseph; sard, Reuben; emerald, Judah; topaz, Dan; beryl, Asher.

Then there is a list given in the description of the ornaments of the Prince of Tyrus (Ezekiel xxviii., 13): 1, *Odem;* 2, *Pitdah;* 3, *Yahalom;* 4, *Tarshish;* 5, *Shoham;* 6, *Yashpeh;* 7, *Sappir;* 8, *Nofek;* 9, *Bareketh.*

In the description of the Heavenly City, (Revelations xxi., 19, 20), another list is given; in this list, which follows, the word used in the original, or Septuagint, is followed by the rendering given by most authorities, that of the Revised Version in parentheses: 1, *Jaspis,* jasper; 2, *Sapfeiros,* sapphire or lapis lazuli; 3, *Chalkedon,* chalcedony; 4, *Smaragdos,* smaragd (emerald); 5, *Sardonyx,* sardonyx; 6, *Sardios,* sardius; 7, *Chrysolithos,* chrysolite; 8, *Beryllos,* beryl; 9, *Topazion,* topaz; 10, *Chrysoprasos,* chrysoprase; 11, *Hyakinthos,* jacinth (sapphire); 12, *Amethystos,* amethyst.

Other references to gems in the Bible indicate the diamond as *shamir,* amber as *hashmal,* and crystal (quartz) as *gerah* and *gabish,* amethyst as *ahlamah,* and it is thought that the pearl is meant by the Hebrew word *peninim,* a word used several times in both the Old and New Testaments as a metaphor for something valuable and precious.

Many and various powers have been ascribed by man to gems; powers curative, talismanic, and supernatural (the word *lithomancy* meaning divination by stones); some gems could be made prophetic, others revealed the past; in the realm of medicine some were prophylactic and most of them were believed to be potent remedies. The latter superstition is hard to kill in the slow dissemination of science, and survives to-day, even in civilised and Christian countries. Some gems were believed to possess the virtue of procuring the favour of the wise or great for their owners, some were supposed to invest their possessors with wisdom, strength, or courage, and some were shields against danger, disease, and death. Gems were connected with astrology, and exerted an influence for good or for evil through the planetary influence of certain days. White stones, the diamond excepted, were to be worn on Monday; Tuesday—the day of Mars—was elected for garnets, rubies, and other red stones; Thursday was for amethysts; Friday—the day of Venus—owned the emerald; Saturday—Saturn's day—claimed the diamond;

while the topaz and yellow gems were appropriate to Sunday.

Particular gems were influential during certain months, and, under the proper astrological control, were supposed to have a mystical influence over the twelve parts of the human anatomy. The potency of a gem worn with regard to this belief was increased if the natal day of the wearer corresponded with its particular sign, and when worn as a birth or month stone was supposed to attract propitious influences and ward off evil. Gems to which were ascribed zodiacal control, and their periods of influence, follow:

Garnet, Aquarius; January 21st to February 21st. Amethyst, Pisces; February 21st to March 21st. Bloodstone, Aries; March 21st to April 20th. Sapphire, Taurus; April 20th to May 21st. Agate, Gemini; May 21st to June 21st. Emerald, Cancer; June 21st to July 22d. Onyx, Leo; July 22d to August 22d. Carnelian, Virgo; August 22d to September 22d. Chrysolite, Libra; September 22d to October 23d. Aquamarine, Scorpio; October 23d to November 21st. Topaz, Saggitarius; November 21st to December 21st. Ruby, Capricorn; December 21st to January 21st.

An idea somewhat similar was that of the Jewish cabalists, which accorded to twelve gem stones, when each was engraved with an anagram of the name of God, a mystical influence with, and a prohetical relation to, the twelve angels, as follows: ruby, Malchediel; topaz, Asmodel; carbuncle (garnet), Ambriel; emerald, Muriel; sapphire, Herchel; diamond, Humatiel; jacinth, Zuriel; agate, Barbiel; amethyst, Adnachiel; beryl, Humiel; onyx, Gabriel; jasper, Barchiel.

The Twelve Apostles were symbolically represented by precious stones: St. Peter, jasper; St. Andrew, sapphire; St. James, chalcedony; St. John, emerald; St. Philip, sardonyx; St. Matthew, amethyst; St. Thomas, beryl; St. Thaddeus, chrysoprase; St. James the Less, topaz; St. Simeon, hyacinth; St. Matthias, chrysolite; St. Bartholomew, carnelian.

While there are variations, the generally accepted list of "birthstones" is:

January, garnet; February, amethyst; March, bloodstone; April, sapphire; May, emerald; June, agate; July, ruby; August, sardonyx; September, chrysolite; October, opal; November, topaz; December, turquoise.

A suggestion of the superstitions which have

invested gems with supernatural qualities follows:

Agate.—Emblem of health and wealth; inimical to venomous things; alleviates thirst; gains victory for its possessor; stays storms; sharpens sight; increases strength; and—a quality that should make it welcome to orators and lecturers—renders its wearers gracious and eloquent. The Mohammedans believed it would cure insanity when powdered and administered with water or apple juice.

Pierre de Boniface, writing in 1315, said:

"The agate of India or Crete renders its possessor eloquent and prudent, amiable and agreeable."

Dioscorides, in his *Materia Medica*, prescribes agate as a preventive of contagion.

Amber was believed to be good for stomach-ache, fits, scrofula, and jaundice. The amethyst—emblematic of sincerity—lost its colour in contact with poisons, and was an antidote for them. It dispelled sleep, sharpened the wits, and promoted chastity; while being a sure preventive of intoxication. Beryl was the favourite stone for divination; reinforced with potent incantations, it foretold the future and reviewed the past. The bloodstone, if rubbed with the juice of the heliotrope, rendered its wearer invisible; it was also a specific for dyspepsia. Carnelian cured tumors, cleared the voice, and preserved harmony; it also stopped bleeding at the nose. Cat's-eye cured croup and colic—it should thus be highly favoured as a stone to be mounted in infant's rings. Chalcedony prevented and cured melancholy; worn in contact with the hairs of an ass it prevented danger during tempests. Chrysoberyl alleviated asthma. Chrysoprase was good for gout. Coral was a fever cure, and has had innumerable curative and preventive qualities ascribed to it. The qualities ascribed to the diamond included the power of curing insanity; powdered it was an excellent dentifrice and it cured epilepsy. In Burma, and in the Middle Ages in Europe, the diamond was supposed to be a poison akin to arsenic. The emerald stopped hemorrhages; it was cooling in fevers and used to strengthen and preserve the eyes. The garnet averted plague and was a defence against thunder, before lightning was known to be the agent of destruction. Jade everywhere and always has rested strong in superstition as a cure for diseases of the kidneys. Jasper was good for lung troubles, was a charm against scorpions and spiders, and would save its wearer from drowning. Jet cured snake bites. Lapis-lazuli cured biliousness. Onyx caused nightmare. Opal was used as an eyestone and heart-stimulant. The pearl cured stomach troubles and skin diseases. Quartz, even to-day, and in the United States, is invested with medicinal and supernatural qualties that hold the firm faith of many persons, especially in remote country places. A "vital ore," which is merely quartz sand, has a vogue in some sections of northern New York State—according to Wirt Tassin—as a panacea; is particularly advocated for sore eyes, hemorrhoids, carbuncles, indigestion, sore throat, giddiness, and blood poisoning. Quartz balls are and have been used with great profit by mystics, astrologers, diviners, and other like fakirs, to foretell the future, disclose the past, and conjure up distant scenes. The ruby is an amulet against plague, poison, sadness, and sensuality; its corundum congener, the sapphire, if placed on the heart, imparts strength and energy; it also cures boils, carbuncles, headache, and cramps. Topaz averts sudden death. The wearer of a turquoise will require no accident insurance, the stone having that power. Zircon stimulates the appetite, aids digestion, and takes away sin.

In India the mystics of that occult land believe in the virtues and malign influences of precious stones; the modern Western spiritualists, who draw upon the Oriental treasure-house of occultism, are said to give credit to gems for mystical properties and influences. A school in Paris teaches a "science" of magnetic emanations, radiance, and crystals, and a Dr. de Lignieres of Nice, France, is the author of a book, in which he seriously considers the medicinal properties and influences of precious stones.

CHAPTER XXIX

FAVOURITE GEMS OF DISTINGUISHED PEOPLE

SENTIMENT occupies a high place in the values of gems, and it has been, to a considerable extent, created by the historical or traditional association of different gems with royal personages and people otherwise famous; the favour of the great has sometimes had an important effect upon the market value of precious stones, and in some cases good or ill fortune has passed with gems from one possessor to another, until to the inanimate jewel has attached the credit or discredit of causing remarkable human experiences, and the stone has acquired the attribute of lucky or unlucky. The diamond fills the leading rôle in this historical and legendary drama of the gems, and a full account of all pertaining to it that is worthy of notice, that is extant in print, might suffice for a volume of considerable interest.

Charlemagne fastened his mantle with a clasp set with diamonds; these historic stones illustrate the crude efforts of the lapidaries of their time, the natural planes of the octahedron being only partly polished.

Louis Duke of Anjou possessed a regal array of jewels; in an inventory of his gems exhibited 1360–1368 was a description of eight diamonds which showed some skill on the part of their cutters.

When the Duke of Burgundy, in 1407, gave a magnificent banquet to the King of France and his Court, the noble guests received as souvenirs of the entertainment eleven diamonds, cut with as much skill as the art of that day was capable of, and set in gold.

Pope Sixtus IV. was the recipient of the second diamond sent to be cut, in 1475, by Charles the Bold, Duke of Burgundy, to Louis de Berquem of Bruges—regarded by his contemporaries as the father of diamond-cutting. The first of the trio of famous stones is said to have been the historic "Beau Sancy"; the third diamond was presented to Louis XI. of France.

"The Twelve Mazarins" were the twelve thickest diamonds of the French crown jewels, ordered by Cardinal Mazarin to be recut by Parisian cutters.

Pope Julius II., in 1500, owned a diamond on which was engraved the figure of a friar by one Ambrosius Caradossa; this is one of the few noted examples of diamond sculpture.

The first French woman to lead fashion as a wearer of diamonds for personal ornaments is said to have been Agnes Sorrel, famous in the time of Charles VII. Subsequently, under Francis I., extravagance in this particular in French society reached its climax, and the Luxus or Sumptuary Laws, in the reign of Charles IX. and Henry IV., were drafted to repress this form of extravagance.

The late Earl Dudley owned one of the several large and world-famous diamonds emanating from the diamond mines of South Africa; this stone was first famous as "The Star of South Africa"; it was then the size of a small walnut, when in the rough, and weighed 83½ carats; cutting reduced it to 46½ carats.

The melodrama of gem history is contributed to by the record of Mohammed Ghori, the real founder of the Mohammedan dominion in India, whose death discovered in his treasury precious stones weighing four hundred pounds, including a great number of diamonds of vast but inestimable value; this hoard of mineral wealth, this enterprising disciple of Mahomet, it is said, acquired exclusively by plunder.

The famous "Eugenie" diamond purchased by the Emperor of the French, Napoleon III., was found by a poor peasant at Wajra Karur in India; the finder tendered the stone to the village blacksmith as compensation for repairing a plough; the smith threw it away, but upon reconsidering its possibilities recovered it and sold it for 6000 rupees to a merchant named Arathoon of Madras, who sold it to the French emperor for a great sum.

Señor S. I. Habid, a wealthy Spaniard of the rue Lafitte, Paris, proprietor of a collection of rare gems, is, according to information published in European and American newspapers during the spring of 1908, the possessor of the famous blue "Hope" diamond. For some time this celebrated stone was owned in America, the possessors being the firm of jewellers in New York City, Messrs. Joseph Frankel's Sons. The American owners admitted the sale of the stone in Paris, but declined to divulge the facts as to the price or the identity of the purchaser, stating that the information, if made public, must come from the purchaser. The Sultan of Turkey was for a time the reputed buyer. Mr.

Edwin W. Streeter, who, partly by virtue of his authorship of *The Great Diamonds of the World*, is entitled to the distinction of the expert on this phase of precious stones, in his book *Precious Stones and Gems*, in a chapter entitled "Coloured Diamonds," traces a complete history of the "Hope" blue diamond. This author is inclined to identify this stone as a part of a blue diamond, bought in 1642 by Tavernier, the famous traveller and gem buyer, supposed to have been found in the old Indian mines, probably those of Gani-Color. It weighed in the rough 112¼ carats; and in 1668 it was sold to Louis XIV. The present name of this diamond is derived from that of Mr. Henry Thomas Hope, a London banker, who bought it in 1830 for the equivalent in currency of the United States of about $85,000.

Among the notable coloured diamonds is the "Dresden green diamond," a fine flawless stone, of a bright apple-green colour. It is in the famous "Green Vaults" of Dresden, and has belonged to the Saxon crown since 1753. Augustus the Strong paid $60,000 for it. Forty carats is its weight.

Another famous forty-carat stone is the "Polar Star," a pure and brilliant diamond, the property of the Princess Yassopouff; it was purchased, prior to its present ownership, by the Emperor Paul of Russia for a large sum.

The Shah of Persia, whose reign has been lately troubled by revolting radicals in his domain, may find consolation in the possession of a vast treasure of jewels rare. These include two magnificent rose-cut stones, the "Darya-i-nur," or "Sea of Light," which weighs 186 carats, and the "Taj-e-mah," or "Crown of the Moon," weighing 146 carats.

The women sovereigns of Austria, beginning with the Empress Maria Theresa, have had the proud privilege of displaying among the crown jewels of the royal house of Austria the famous "Florentine" diamond, also known as the "Austrian Yellow" and the "Tuscan" diamond. This illustrious citron-yellow stone weighs 139½ carats and is cut into a nine-rayed star of the rose form. The "Florentine" was formerly owned by the Grand Duke of Tuscany.

CHAPTER XXX

GEM MINERALS AND GEMS IN MUSEUM COLLECTIONS

VISUAL and palpable examination of gems and gem minerals is most desirable, if one would have a thorough understanding of gemology, for all that the best of books can teach must necessarily be, to a considerable extent, abstract. Fortunately for those who abide or sojourn near enough to take advantage of them, there are several public museums in America which possess collections of minerals, including gem minerals, and in New York City the great educational institution, The American Museum of Natural History, has, in addition, a fine collection of cut gems, principally the gift of Mr. J. Pierpont Morgan, which is a delight to the eye of every visitor who sees it. While one cannot handle the minerals in such collections, and thus prove the statements made in this book and other publications, that gems are cold and that some feel greasy or have other qualities determined by the tactile sense, they are free for all to study optically, and so plain and practical is their scientific and common-sense arrangement, that the appreciative student must feel in his heart a great sense of thankfulness, not only to the generous men of wealth, who by gifts and endowments have created this magnificent institution, but also to the curators who have by their arrangements in exhibiting and labelling, with the auxiliaries of "rubrics" and guides and other publications, made the study of these representative specimens of minerals so easy that it might almost be said that "he who runs may read." The students of gems in New York owe to the generosity of Mr. Morgan the two large Tiffany exhibits of precious stones which were prepared by Tiffany & Co., under the direction of Dr. George Frederic Kunz, and exhibited, with distinction and credit, at the Universal Expositions of 1889 and 1900 at Paris. These two collections are now incorporated in the general exhibit of gems in the Gem Room at the museum. In connection with these exhibits, and as a recognition of his public services in behalf of art and science, Mr. Morgan was made by the French Republic *Officier de Legion* d'Honneur. Mr. Morgan also presented to the museum the superb mineralogical collection of Mr. Clarence S. Bement, of Philadelphia, which has for years stood foremost among American cabinets, and vies (especially in the matter of American minerals) with the great collections of the world. In this connection it is interesting and appropriate to record the generous gift of Mrs. Matilda W. Bruce of New York City, who created the Bruce Fund; this is an endowment, of the sum of ten thousand dollars, of the Department of Mineralogy of the American Museum of Natural History, which yields an annual income of $660, which is applied to the purchase of specimens. The development of minerals is the slowest growth in the scheme of creation, but it is a satisfaction to know that in the American Museum of Natural History, as in other "live" kindred institutions, the collection of minerals develops and improves rapidly, as is well known to those who have solicitously kept pace with it year by year. For the student who would go deeper than to the extent of a mere fancy, there exist associations most helpful and interesting, of which the student can be the beneficiary and a member at very slight cost; such as the New York Mineralogical Club and the Philadelphia Mineralogical Club, which hold educative meetings where the members read papers and in many ways contribute information, and which make field study trips to localities known to be productive of specimens of interest. All who visit the collections at the American Museum of Natural History should obtain *Guide Leaflet No. 4 for the Collection of Minerals* (which is a supplement to the *American Museum Journal*), written by Louis P. Gratacap, A.M., Curator, Department of Mineralogy, of the museum. For more extensive information applicable to this collection and institution, and to similar ones, a most profitable investment would be the book by the same author, *A Vade Mecum Guide to Mineral Collections, with a Chapter on the Development of Mineralogy*, with enlightening half-tone illustrations and over two hundred figures of crystals. There are also periodical publications devoted entirely or in part to mineralogy.

The growth of the mineral collection of the American Museum of Natural History has been gradual, beginning with the Bailey collection, which served as an introductory and fairly representative series of specimens. A valuable accession was the most remarkable group of specimens of malachite and azurite donated by the Copper Queen Consolidated Mining Company of Arizona, which, with subsequent additions from the same donor, is the most striking feature of the whole collection; it is assembled and installed in a single case at the north end of the small hall. After this invaluable acquisition of the green and blue carbonates of copper from Arizona, the Spang collection was purchased in the year 1891, which doubled the number of specimens possessed by the museum, and added many new varieties and kinds of minerals. In the nine years that followed many valuable additions came from generous benefactors, and in 1900 Mr. J. Pierpont Morgan purchased and presented to the museum the remarkable collection assembled by Mr. Clarence S. Bement of Philadelphia, characterised by the collector's superior scientific judgment and exquisite taste —which evolved from the field of specimens available throughout the world a great variety of forms representing the commoner minerals—and the exceptional perfection of the specimens.

While the rock-bottom upon which modern mineralogy is founded is chemical law, it might be said that crystallography is its foundation, so that minerals of the same chemical type are grouped together, in the modern scheme of exhibition; and, under that type, minerals of similar physical or crystallographic features are arranged in smaller subdivisions. To quote Professor Gratacap:

> The forms of minerals are their most obvious characteristic. The six-sided prisms of quartz and beryl crystals, the rhomboidal or trapezoidal faces of garnet, the triangular faces of magnetite and the square faces of fluorite are unmistakable.

The branch of mineral science known as crystallography is now well developed and established, and it has been demonstrated that crystal form has a close dependence upon chemical composition. The arrangement of all specimens at the American Museum of Natural History, in both desk and wall cases, is exemplarily systematic, and in accordance with the classification of the sixth edition of Dana's *System of Mineralogy*. An intelligent inspection of the collection at this museum, for the novice in mineralogy, should begin with desk case No. 28, followed by case No. 27; these two cases contain introductory series presenting the chemical and physical features of minerals, together with explanatory tables and photographs. The models showing the formation of crystals are ingenious in design and excellent in construction, and illustrate the crystallographic system to the

A CELEBRATED COLLECTION OF GEM MINERALS
Courtesy of Charles F. Wood

novice clearly and as no other device possibly could do. Visitors to the museum who are in the jewelry trade are likely to view with particular interest the choice specimens of gold exhibited in desk case No. 1, where it appears in sheets like rolled metal; in plates, with crystallised edges; in braided filaments made up of minute octahedrons with hollow faces; in twisted plates frequently attached to quartz, around which it curls like some irregular yellow flower; besides which there are cavernous, skeleton, and pitted crystals; peculiar distortions; reticulated and tree-shaped groups with spongy masses; and rounded water-worn nuggets. Case No. 27 also contains the fine collection of the New York Mineralogical Club of specimens of minerals occurring on Manhattan Island; these include garnets, zircon, and tourmalines and a few other gem minerals, although not all of gem quality.

In the south end of the small hall is the collection of gems which, while it is not as broadly representative of the semi-precious stones as it could be, provides an ocular demonstration of the appearance of typical gem minerals of good colour and qualities, advantageously cut. A brief visit to this collection, as a supplement to the study of gems through books, will provide a practical lesson that will clearly illustrate the written descriptions of precious stones, and leave a mental picture that is likely to be lasting.

BROOCH, FESTOON, RING, AND EARRING; SUGGESTIONS FOR STUDENTS AND JEWELLERS

Designed in 1910

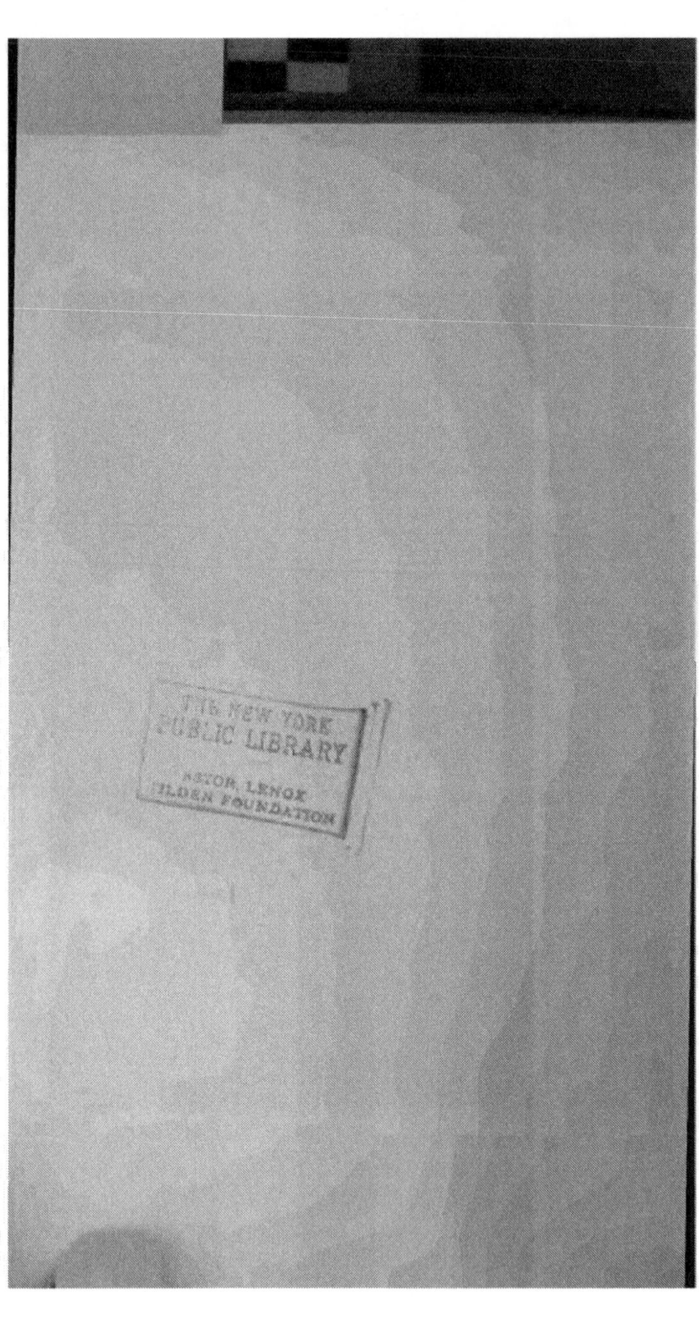

CHAPTER XXXI

OUR DIAMOND CUTTERS AND THEIR TRADE UNION

THE trade of diamond-cutting presents many points of interest, beginning with the high intrinsic value of the raw material entrusted to these workmen, upon whom their employers must rely for absolute honesty, rare skill, and the best of judgment. The diamond cutters in North America are not a great power numerically in the world of labour, but their labour union is in some respects one of the strongest of such organisations.

Peter Goos, the first diamond polisher to settle in the city of Amsterdam, Holland, arrived there in 1588. In time the mere bruting or polishing of diamonds in Holland was succeeded by scientific cutting on geometrical lines and the artisans employed in the work and their processes were evolved into a distinct and recognised industry. In the year 1815 the leading diamond cutters of Holland convened, declared themselves "masters," decided to employ, to begin with, a score of apprentices, and organised diamond-cutting into a full-fledged trade. The foundations being thus laid, the trade flourished until the last half of the nineteenth century, when it apparently was obliterated as one of the effects of war, chiefly the Civil War in the United State and the Franco-Prussian War in Europe. When the first diamond mines were discovered in 1870 in South Africa, the demand for diamonds rose, and diamond cutters were once more enlisted in the service of the Dutch, English, and French importers, and almost any one who wished was given an opportunity to learn the trade, which had been so long asleep. The trade in diamonds then rapidly developed annually; improved steam navigation and other scientific progress provided better facilities for exporting and importing gems, and there were established many new factories for cutting and polishing diamonds in the city of Amsterdam, until the entire industry centred in Holland's capital. Amsterdam only secured the lead as the Diamond City after a keen commercial and industrial rivalry with Antwerp, a contest that was waged, with varying fortunes, for many years. The workmanship of the diamond cutters and polishers of the Amsterdam factories is first

SUGGESTIONS FOR STUDENTS AND JEWELLERS.

Designed in 1910

class and the standard for the trade throughout the world.

The diamond cutters' union of Amsterdam is a trade union of unique solidarity, which has been tried by the fire of many industrial disputes and trials, particularly during dull times when but a portion of the members could find employment. There are at the present time eighty-five hundred workmen, all members of the union, in Amsterdam, distributed among some eighty factories. The Amsterdam union is governed by salaried officers, who are elected by the whole body. These officers are: president, secretary, treasurer, and second treasurer; also an inspector of wages, whose function and duty it is to investigate and report upon any violation of a wage agreement he may discover. The union publishes a weekly journal, edited by the union's president; this journal is regarded by the members of the union as the foremost authority upon all matters connected with the diamond industry. The Amsterdam union was organised in November, 1894, after a simultaneous strike of all the operatives. The strike and union followed a commercial depression of the diamond trade and a consequent reduction of wages. Prior to the discovery of diamonds in South Africa in 1870, the diamond cutters of Amsterdam received an average of from sixteen to eighteen dollars per week; directly after the discovery, when diamonds were found in large quantities, a period known in the trade as "the Cape time," the demand for the skilled labour of the cutters was so great that wages were increased so that the diamond cutters were able to earn from two hundred to six hundred dollars per week; this is a conservative statement, for a diamond cutter now employed in New York City states that his father, employed in Amsterdam during that time, earned as high as eight hundred dollars in one week.

The eighty-five hundred diamond workers of Amsterdam are divided into ten branches, known as follows: No. 1, brilliant polishers; 2, brilliant polishers' assistants or helpers; 3, brilliant cutters; 4, brilliant setters; 5, rose polishers; 6, rose cutters; 7, rose setters; 8, six- and eight-face polishers; 9, cleavers, or splitters; and 10, sawyers. Each of these branches has its own delegation to represent its members in the executive board of the union.

In North America the diamond cutters are well organised.

When the United States levied an import

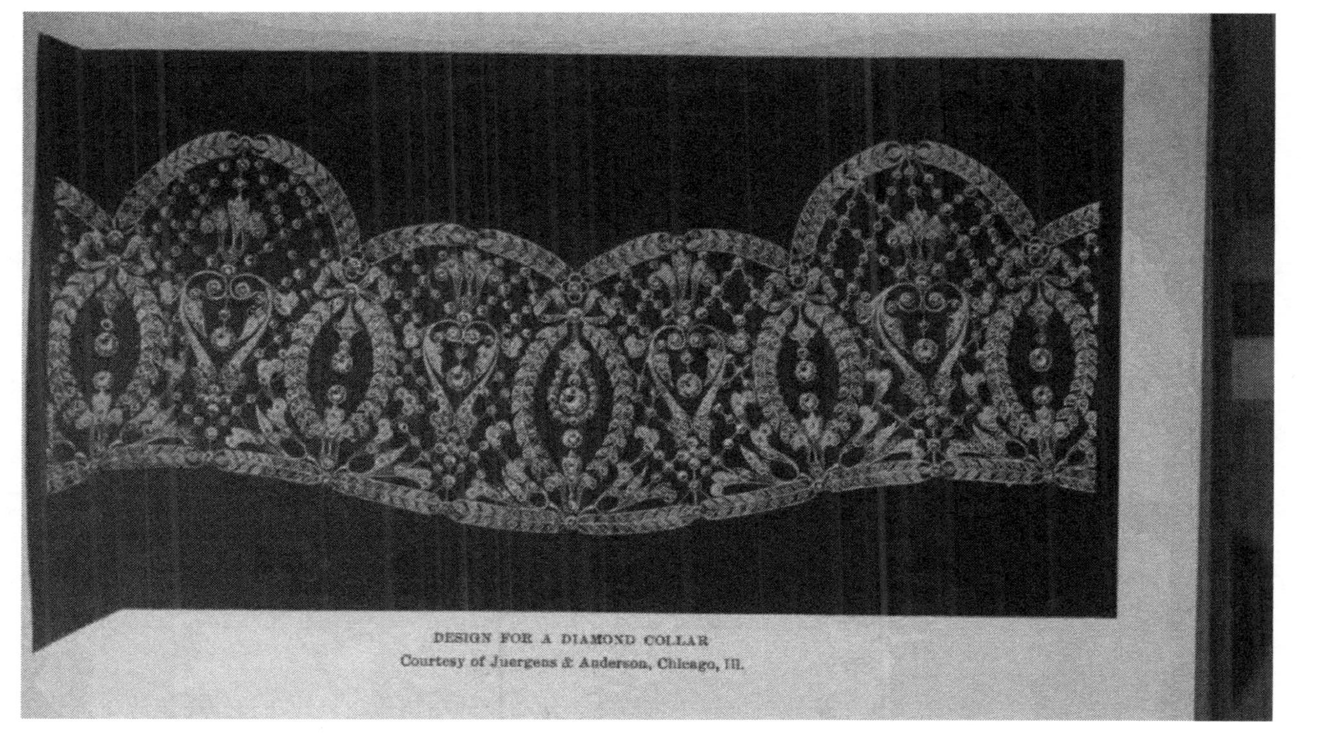

DESIGN FOR A DIAMOND COLLAR
Courtesy of Juergens & Anderson, Chicago, Ill.

duty on diamonds, there arose a demand for expert operatives to cut and polish diamonds here, and then came the first immigrant diamond workers, mostly from Amsterdam. As soon as there was a sufficient number of diamond workers here to form a numerically respectable organisation, which was in 1895, the men established their first union. The Dingley Tariff, which provided a duty of ten per cent. on uncut diamonds and twenty-five per cent. on cut stones, had been enacted into a law, and it profited American importers to have their diamonds cut here, and cut in accordance with the exacting requirements of the American trade; so diamond-cutting was raised into a small but a recognised industry. The first union organised, although successful from its inception, disbanded, because the membership represented too many different nationalities and customs, and the individual members had not then learned the wisdom of subordinating petty prejudices and motives to the common interest.

The present union is entitled The Diamond Workers Protective Union of America, and was organised September 16, 1902. There are about three hundred and seventy-five members, a majority being natives of Amsterdam, although,

mours, France; Geneva and Gex, Switzerland; London, England; and New York. Through this central organisation, all diamond workers of the world are virtually under one control. When a member of one local union goes to another place, he receives a certificate which entitles him to membership in the organisation existing in the place of his destination, and he is entitled to immediately participate in all benefits that the local union may afford. Reports issued monthly by the International Board enable the affiliated local unions to keep track constantly of the conditions of the various markets of the world. The local unions contribute to general strikes in other countries and are assessed, if necessary, so that strikes can be continued after the fund of the local treasury has run out. All news of worthy importance to the workers in the diamond industry is promptly cabled. If a union proposes to change the wages or other conditions, its claim is presented to the individual employers. If employers and employees cannot agree, the differences are usually first referred to the Diamond Cutters' Manufacturers' Association, which in most cases, appoints a committee to arbitrate the questions at issue, with a corresponding committee of the union. From January, 1906, until May, 1908, trade agreements existed between the employers' and employees' associations in the United States, whereby hours of labour, scales of wages, apprentice regulations, and practically all matters which could result in conflicts, were regulated. For matters which were not covered in these agreements a clause provided that recourse must be had to arbitration.

The diamond-cutting industry in the United States was in a flourishing condition from its beginning until the latter part of 1907, when, because of the financial depression popularly termed "the rich man's panic," all the diamond-cutting factories in America were closed, throwing out of employment the entire number of diamond workers. Before the advent of the ensuing year a few factories reopened with work progressing on a small scale, and, gradually, as confidence in the commercial world was restored, the factories resumed operations. During the period of idleness about one hundred of the workmen in the trade returned with their families to Amsterdam and Antwerp, where they received immediate employment.

At the beginning of the panic of 1907 the American diamond cutters' union had a surplus

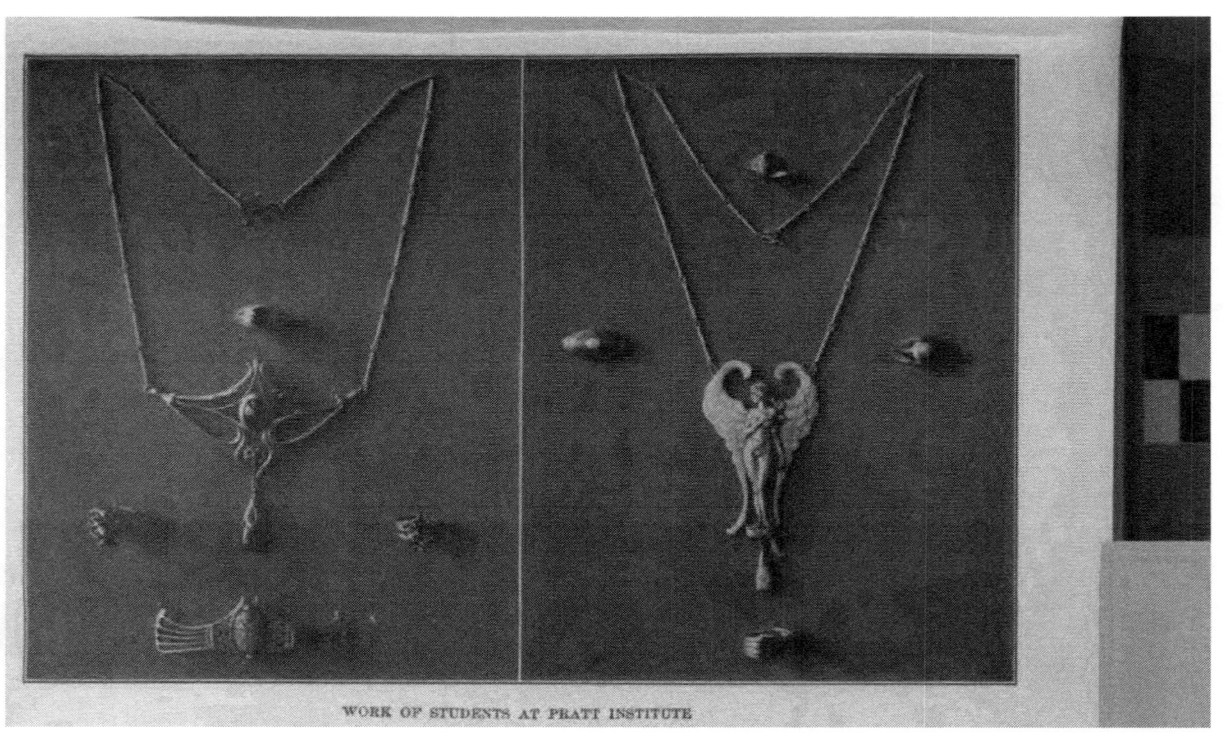

WORK OF STUDENTS AT PRATT INSTITUTE

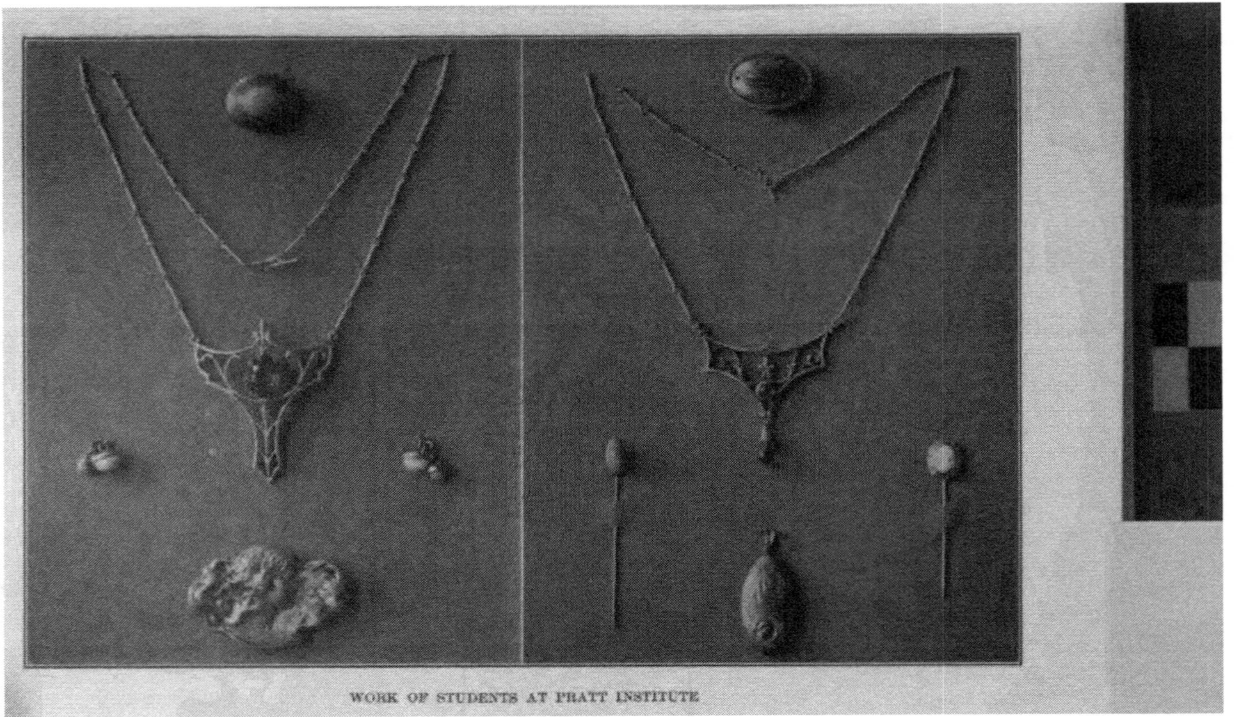

WORK OF STUDENTS AT PRATT INSTITUTE

in its treasury of $27,000; this sum was soon used up for the support of members, and the union in Amsterdam remitted a maintenance fund of $15,000.

CHAPTER XXXII

JEWELRY IN THE ARTS AND CRAFTS MOVEMENT

THE sequence to the cutting of a gem is generally mounting and setting it, unless it is merely perforated and strung as a bead or hung as a pendant. Mounting and setting is the trade of the goldsmith or jeweller, and whether his goods are artistic or inartistic depends to a great degree upon the discrimination of buyers. There is almost as much variation in the metallic environment of gems as there is in architecture, and the designing and execution of the jeweller range from meritorious to atrocious. To a great extent the metal mountings for gems are stamped out in dies or are otherwise machine-made, but no matter how deserving of praise the original design, the finished article, to the eye artistic, is "commercial." Within a few recent years the struggle to elevate art, in other directions than in the field of things considered as exclusively its province, has invaded the domain of jewelry, and some patient work-

SUGGESTIONS FOR STUDENTS AND JEWELLERS

Designed in 1910

ers have produced commendable creations by their handicraft. This new jewelry is partly identified with what might be termed the general arts and crafts movement, but, as is always the case with efforts of this kind that become known under a popular name, many unworthy deeds are done under its banner by the careless, the deceptive, or the undisciplined, whose products, heralded by them as "artistic," are worse than "commercial." Pretenders can easily impose upon the uneducated. But honest efforts are being made by pioneers with high ideals to properly instill them into the minds of student craftsmen, and to train their hands to a degree of skill that will measure up to the higher standard, which hopeful reformers are trying to set for the jewelry of the future. The efforts of these idealists of the arts and crafts movement deserve the respect, the encouragement, and the co-operation of gem dealers and of the jewelry trade throughout. As it has been well said by Professor Oliver Cummings Farrington in his *Gems and Gem Minerals:*

There is room, however, for the development of a much higher taste in these matters than exists at present. The average buyer is content to know that the article which he purchases contains a sapphire, emerald, or diamond, representing so much intrinsic value, without considering whether the best use of it, from an artistic point of view, has been made; or whether for the same outlay much more pleasing effects might not have been obtained from other stones. In the grouping of gems, with regard to effects of colour, lustre, texture, etc. certain combinations often seen are far from ideal, while others rarely seen would be admirable. Thus a combination of the diamond and turquoise is not a proper one, since the opacity of the latter stone deadens the lustre of the former. The cat's-eye and diamond make a better combination, and so do the more familiar diamond and pearl. Colourless stones, such as the diamond or topaz, associate well with deep-coloured ones, such as amethyst and tourmaline, each serving to give light and tone to the other. Diamond and opal as a rule detract from each other when in combination, since each depends upon "fire" for its attractiveness.

While there are variations innumerable of design and device in mounting gems, there are practically but two basic methods, the mount *à jour* (two French words, meaning *to the light*) and the encased mount. The ordinary manner of setting gems in rings, the stone held by a circlet of claws, permitting a view of it, or through it, from all points, illustrates the

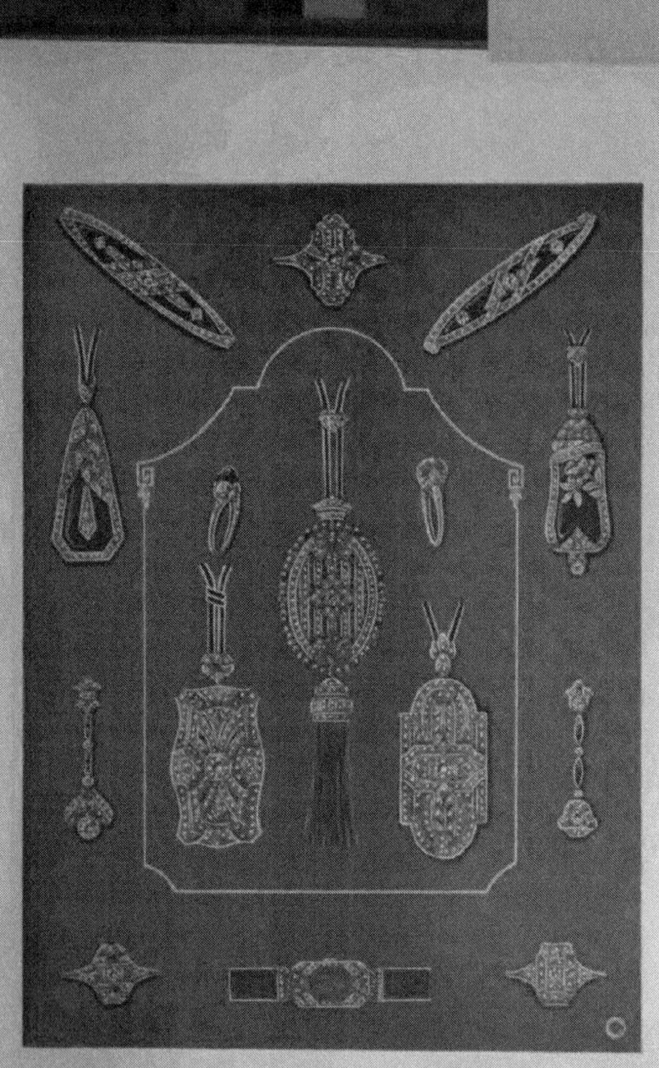

Courtesy of "Gems Creations"

STYLES OF PLATINUM DIAMOND JEWELRY, 1924

Designs by S. Kaufman

à jour, or open, method. This is best adapted to transparent stones, exposing them freely to the light. The projecting claws of the open setting are slightly cleft near their extremities and these, under a pressure that inclines them slightly inward, pinch or grasp the stone at the girdle. Opaque stones, such as turquoise, bloodstone, or onyx, are usually set in the encased mount, in which the gem is set in a metal bed, with only the top exposed.

While to some degree anything fashioned by machinery is open to the detracting term " commercial," there is often much artistic merit in the designs issuing from the factories of manufacturing jewellers, but nothing can rival the charm of objects wrought solely and entirely by hand.

The work of the more expert of the students taking the jewelry course in Pratt Institute of Brooklyn, and at other educational institutions where this department of art and manual training is a serious feature, is a revelation of present attainments, and a hopeful sign that the jewelry of the future in America will conform more to true artistic ideals and serve less as a medium for mere ostentatious display. An exhibition of the work of students in the jewelry course was an attractive phase of the twenty-fifth annual exhibit of student products at Pratt Institute in June, 1908. The exhibits of the class in jewelry and metal-chasing were displayed in two large glass cases, and consisted of rings, pendants, bracelets, stick-pins, brooches, scarf-pins, buckles, and hammered copper work.

A silver medal presented by Mr. Albert M. Kohn of New York City, as a prize for the most proficient student of the jewelry class, was awarded by a committee of trustees, who acted as a jury of award, to Mr. Carl H. Johonnot. The work exhibited by the winner of the medal included a number of fine pendants, rings, silver spoon, and stick pins.

For a description of the class in jewelry designing at the Pratt Institute, and also for excellent photographs of finished work executed and designed by students of the class of 1908, credit is given to Mr. Walter Scott Perry, Director of the Department of Fine and Applied Arts, of the Institute.

The first class in jewelry, hammered metal, and enamelling was organised in the Department of Fine and Applied Arts, Pratt Institute, in September, 1900, with Mr. Joseph Aranyi as instructor in day and evening classes. Mr.

PRIZE DESIGN: WORK OF FREDERICK E. BAUER, STUDENT, JEWELRY CLASS, COOPER UNION

Aranyi at the time was one of the expert workers with Messrs. Tiffany & Company, New York City. He continued as instructor of the class, until June, 1904, when he resigned his position to accept one in Providence, Rhode Island.

In September of the same year Mr. Carl T. Hamann was appointed instructor in jewelry, and for some time has had full charge of all work of this class. He has proved himself an exceptionally fine instructor, and the quality of work has gained very rapidly under his instruction. Mr. Hamann is an expert jeweller by profession, being formerly connected with Durand & Company, Newark, N. J., and later with Tiffany & Company, New York. In 1889 he went to Europe and studied modelling in Munich for one and a half years, going from there to Paris, where he studied in the Académie Julian and the École des Beaux Arts for two years. After his return to this country he became the head modeller for the Whiting Manufacturing Company, New York. Mr. Hamann was the sculptor of the statue of Justice which was one of the eight statues on the Triumphal Bridge at the Pan-American Exposition at Buffalo. At St. Louis he had a statue symbol-

studios of their own and fill orders that come to them from many and varied sources.

The courses are planned to meet the needs of those who wish to enter the trades involving jewelry, enamelling, repoussé, chasing in precious and other metals, and the making of suitable tools required in such work. They give adequate training in design and modelling, in the application of designs to practical problems, the setting of stones, enamelling and finishing, and in the methods and practice of technical work in metal. Instruction is also given in medal work and in the preparation of models for reduction.

The increasing demand for applied art work in useful objects, and the difficulty experienced by manufacturers in securing the services of American artisans whose knowledge and skill are sufficient to guarantee good workmanship, present a trade condition which offers unusual opportunities for remunerative employment and advancement to those who have had the advantage of such training as these courses give.

In this day of specialisation, the apprenticeship system is no longer adequate. The apprentice acquires little more than the skill necessary to meet the technical requirements of his trade;

but, as the success of the ornamental metal worker depends quite as much upon his artistic conceptions and his designs as upon his skill in execution, the work of the shop must be supplemented by art instruction. By alternating the character of the problems given to the students, the applied work shows the inspiration that comes from a careful study of modelling, ornament, and the principles of design; and the work in modelling and design shows the adjustment and illumination that come from constant contact with practical problems.

The courses appeal to two classes of workers; to the apprentice who, by this instruction, can greatly shorten the period of his apprenticeship, and who can supplement the technical skill which he would gain in the shop by the work in drawing, modelling, and design; also to the art student who is turning his attention to work in the applied arts. The opening offered to such a man in this field exceeds that in almost any line of illustrative art work; and the demand for trained workers in the skilled trades in art applied to metals and the limited supply of such men make advancement practically assured to an earnest worker.

The rooms of the department devoted to the

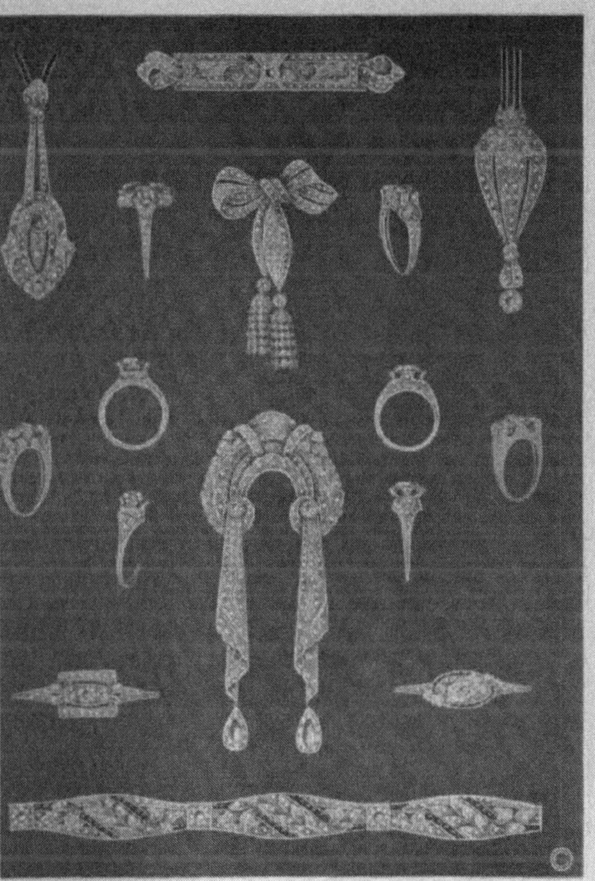

Courtesy of "Gems Creations"

STYLES OF PLATINUM DIAMOND JEWELRY, 1924

Designs by S. Kaufman

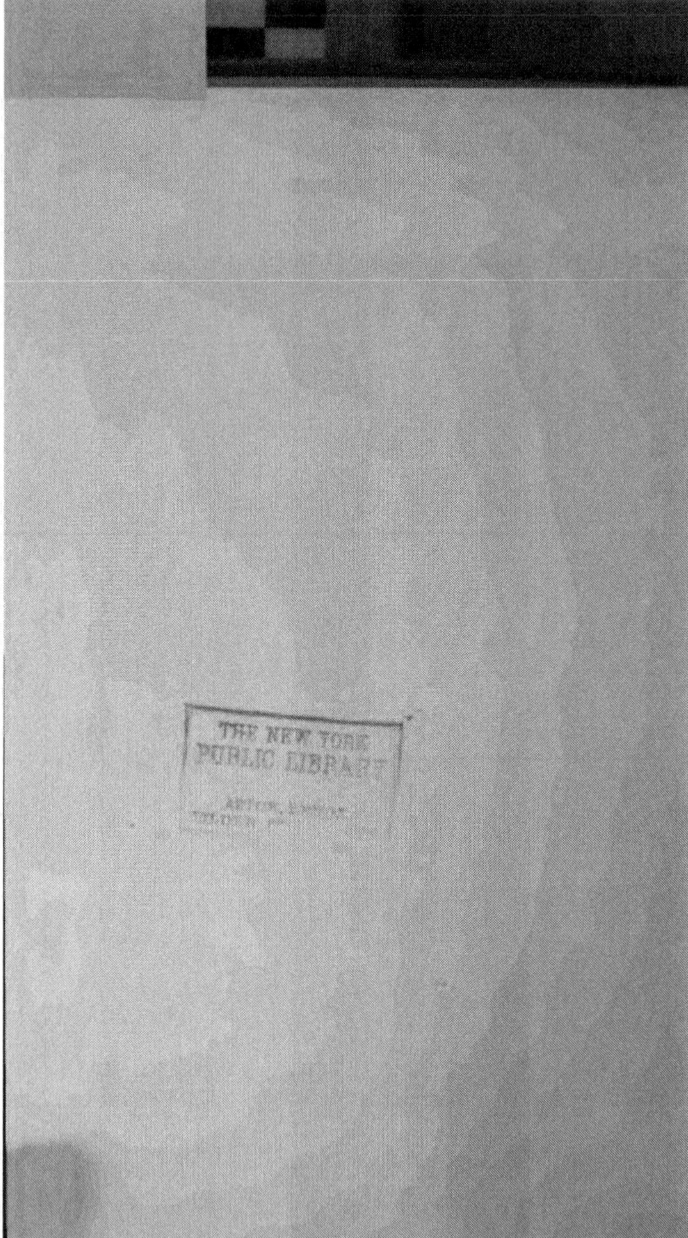
THE NEW YORK
PUBLIC LIBRARY

study and practice of jewelry and other forms of metal work are equipped as workshops with everything needful for practical and applied work.

The day course includes instruction in drawing, design, historic ornament, and in applied work in chasing and repoussé, jewelry, enamelling, and medal work.

All work is designed and modelled in wax, cast in plaster, and then wrought in copper, silver, or gold. In the work in jewelry, silver is used from the first, students making rings with various stone settings, scarf pins, pendants, chains, bracelets, buttons, brooches, etc., the work being plain, decorated, chased, or set with stones.

In hammered metal work, students make their own tools and produce shallow and deep objects in copper and silver, including trays, bowls, spoons, and the like, with decorative designs and repoussé chasing. Parts of objects, such as handles and supports, are also cast, chased, and applied as needed in the design.

Instruction is given in enamelling on copper, silver, and gold.

All work is done in a thoroughly professional manner. Applicants are accepted only for regular and systematic work, and they must give evidence of originality, skill, and general fitness for the course.

Certificates will be granted for the satisfactory completion of a day course of three years.

The classes meet for work daily, except Saturday, from 9.00 A.M. to 4.25 P.M. Instruction is given on eight of the ten half-day sessions. The tuition fees are, $20.00 a term, with an additional laboratory fee of $3.00 a term for miscellaneous material used by students. There are three terms in each school year. The fall term opens the last week in September.

The course provides for wax-modelling, hammered metal work, the application of relief ornament, and the finishing of casting in a thoroughly professional manner; the work being planned for advanced students as well as for beginners. Instruction is given in the making of tools, the modelling of objects in sheet metal, repoussé, or relief ornament in flat and hollow ware, and the chasing of ornament in brass, bronze, silver, and gold. Instruction is also given in jewelry. The class meets on Monday, Wednesday, and Friday, from 7.30 to 9.30 P.M., from the last week of September to the last of March. The tuition fee for the evening course

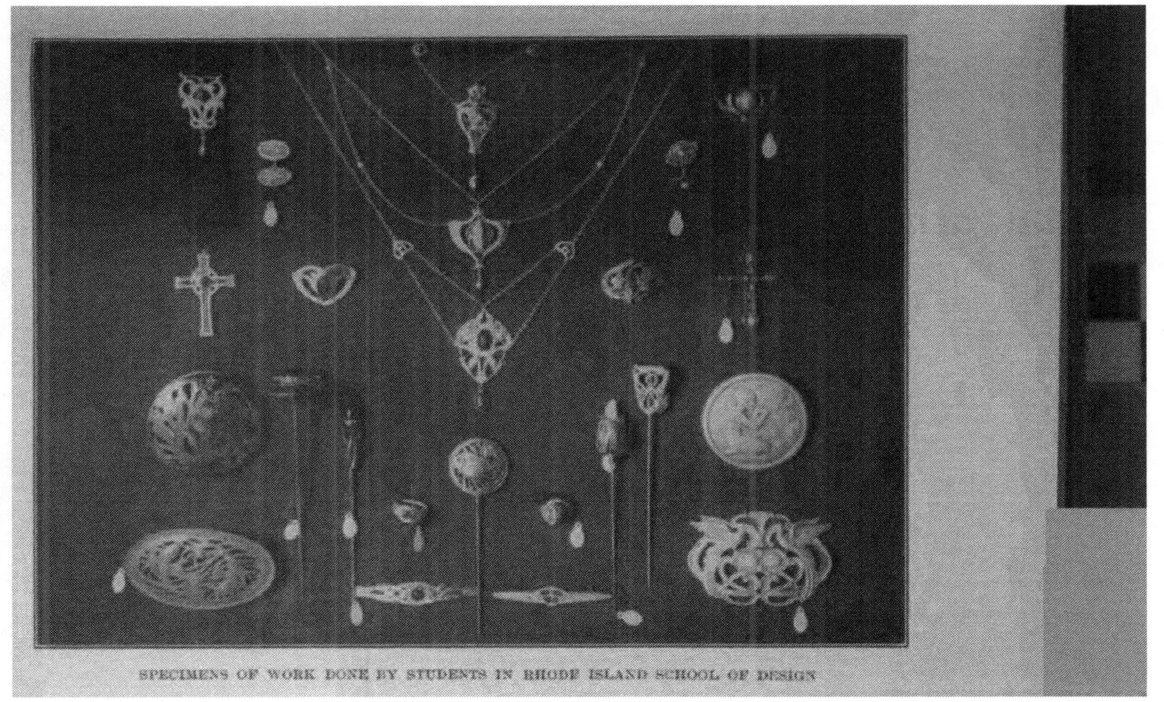

SPECIMENS OF WORK DONE BY STUDENTS IN RHODE ISLAND SCHOOL OF DESIGN

is $15.00 a season of six months, which includes all practice material used by students in class.

Students and alumni of Pratt Institute have organised The Pratt Art Club, which its members otherwise quaintly designate as "Ye Brooklyn Club of Ye Handicrafters"; in its exhibitions, held at the club's rooms near Pratt Institute, are shown some attractive specimens of the work of these crafters.

There is a course in jewelry designing at Cooper Union in New York City under the direction of Mr. Edward Ehrle. The Cooper Union class meets tri-weekly, in the evening, for a two hours' session. The work begins with easy geometrical designs; original work by the pupils is constantly encouraged. The school year begins the second Monday after September 15th and ends about May 15th. The full course requires about three years. At the conclusion of the term in the year 1908, a cash prize offered by *The Jewellers' Circular-Weekly* was awarded to Mr. Frederick E. Bauer for his excellent work.

A resource of value to the artistic designer of jewelry in and near New York City is the Cooper Union Museum for the Arts of Decoration, a subsidiary institution of this famous old hall of education that is now, although progressing in its acquisition of valuable exhibits, of incalculable value to the arts and industries of America; the usefulness of this institution is however restricted, because it is not well known. It is probably a safe assumption to say that not one person in many thousands of the inhabitants of the metropolis is cognisant of the existence of such a treasure-house, which is available to all earnest seekers after ideas, information, and material for the betterment of art, and under conditions impossible to excel in providing the greatest opportunity and freedom to all who will avail themselves of it. The contents of this museum would astonish thousands who are familiar with the broadly advertised contents of the Metropolitan Museum of Art, and the feeling of regret that comes over the appreciative visitor to the Cooper Union Museum suggests the reflection that a little adept yet dignified promotion of publicity would be beneficial to the efficiency of this institution. A strong feature of this working museum is a collection of encyclopedic scrap-books, open, like all else here, to all applicants for permission to use them; the scrap-book covering jewelry shows

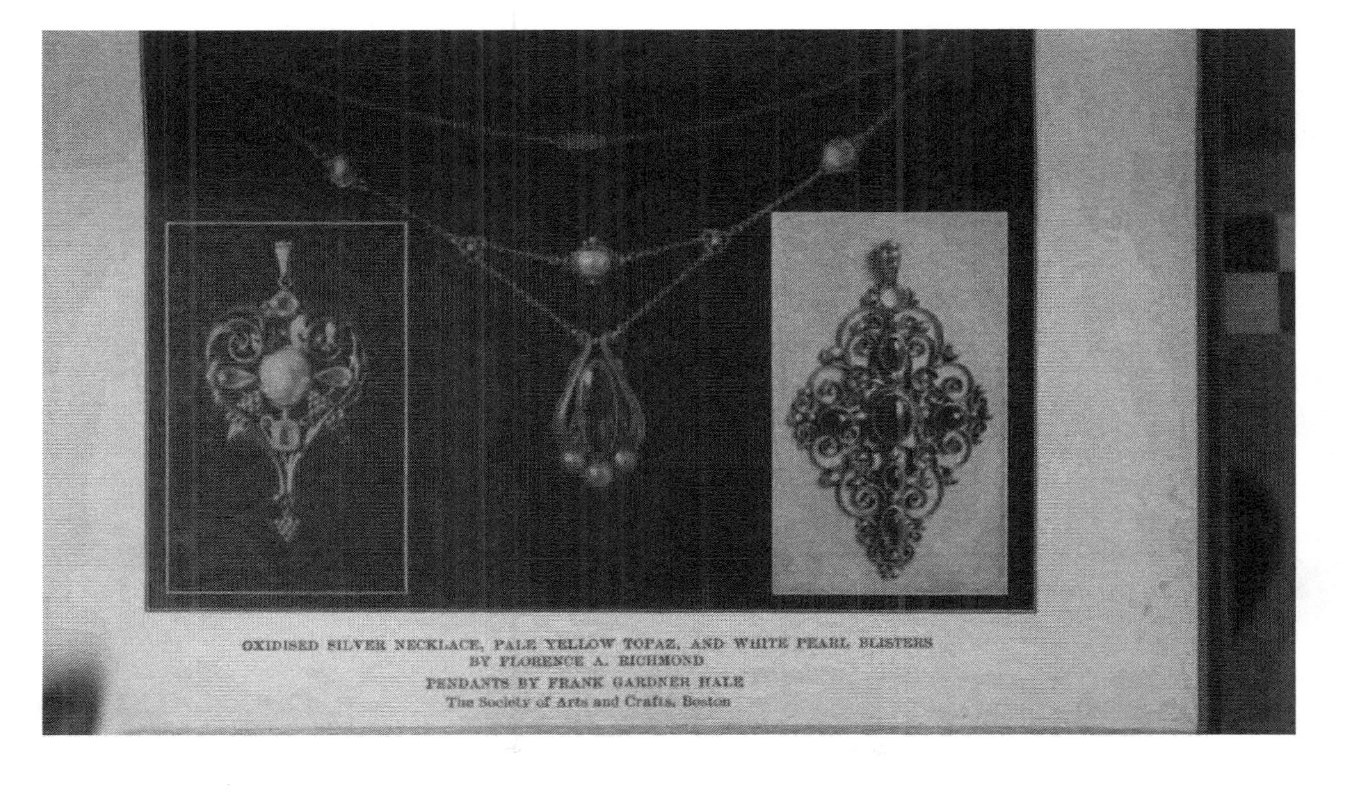

OXIDISED SILVER NECKLACE, PALE YELLOW TOPAZ, AND WHITE PEARL BLISTERS
BY FLORENCE A. RICHMOND

PENDANTS BY FRANK GARDNER HALE
The Society of Arts and Crafts, Boston

many excellent designs, fertile in ideas for bracelets, chatelains, clasps, lockets, combs, crowns, tiaras, head ornaments, dress and engraved ornaments, knots and bowknots, earrings, girdles, belts, hoops, rings, necklaces, pendants, sceptres, seals, and watches.

While the bibliography presented in this volume is extensive and of wide scope, unfortunately, but a few of the books listed are to be found in the average public or institutional library. A valuable resource for the students at Pratt Institute or Cooper Union, or any one who would delve as deeply as possible into the subject of jewelry, is the Society Library in University Place, near Thirteenth Street, New York City. This, Manhattan Island's oldest library, was founded by King George II., and his representative who was at the time the royal governor of the Colony of New York. The family of ex-President Roosevelt have been benefactors of the library for six generations, and he is at this time an active member of the board of trustees. Although not a public library, the superb collection of art books, selected with special reference to the requirements of artists and handicraftsmen, is always open to designers. There is a large endowment fund for the support of the art book department, which is known as "the Greene foundation."

The productions of designers and workers in jewelry seen in the annual exhibitions now held by the National Arts Club, in collaboration with the National Society of Craftsmen, in the galleries of the club at 119 East Nineteenth Street, New York City, prove the good work that is being done by individuals and members of various schools and classes; these include the jewelry class of the New York Evening School, and the jewelry class of Miss Grace Hazen of Gloucester, Mass.

At Newark, N. J., an industrial city which includes among its industries considerable jewelry manufacturing, there is the Newark Technical School, supported by appropriations from both the city of Newark and the State of New Jersey, which has a valuable course for workers in jewelry.

In Boston there is continuous encouragement to designers of art jewelry in the work and influence of the Society of Arts and Crafts, Boston, incorporated in 1897, and which holds exhibitions semi-annually. A recent exhibition of this society included a valuable and most interesting display of American jewelry, the

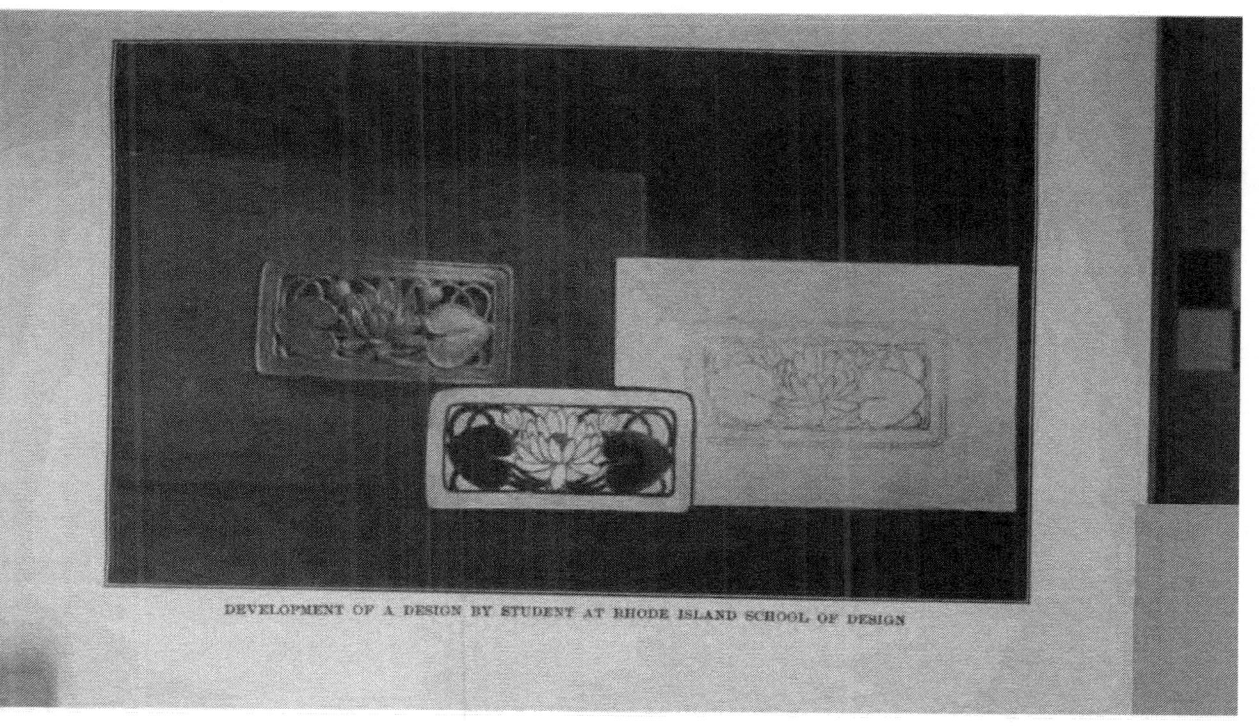

DEVELOPMENT OF A DESIGN BY STUDENT AT RHODE ISLAND SCHOOL OF DESIGN

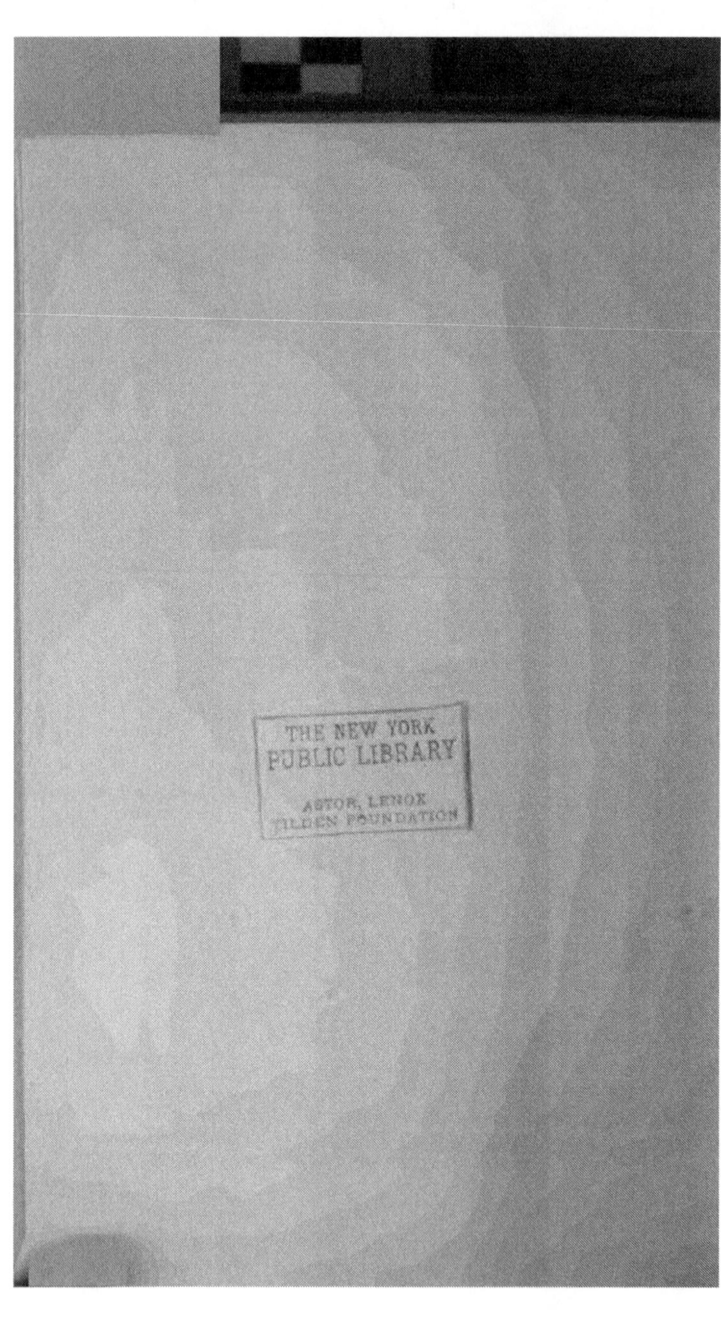

feature of which was a large collection of exquisitely designed, excellently drawn, and well executed pieces from the Copley Square Studio of Frank Gardner Hale, the exhibit occupying one end of the exhibition gallery. Mr. Hale's products are not only definite in design, but the construction of his mountings of gems is practical and would satisfy the mechanical requirements of manufacturers of jewelry commercial, which a good deal of the work of exponents of arts and crafts jewelry would not. New Yorkers at home have had an opportunity to see some of Mr. Hale's remarkable work at an exhibition at the Clausen Galleries. Among the designs exhibited, chains, necklaces, pendants, and brooches predominated; there were numerous crucifixes in silver, some of them containing precious and semi-precious stones. In the number and excellence of these crucifixes, Mountford Hill Smith took the lead among the exhibitors. Marblehead's handicraft shop was represented by the work of H. Gustave Rogers. Commendable work was shown by Jane Carson and Theodora Walcott. Notable exhibits were those of Laura H. Martin, Elizabeth E. Copeland, and Martha Rogers. Ingenious schemes of colour in small enamels were shown by Mabel

W. Luther. William D. Denton of Wellesley exhibited "butterfly jewelry" in which the wings of the butterflies are protected by rock crystals set in gold mounting. Florence A. Richmond and Jessie Lane Burbank from the workshop in Park Square exhibited pieces deserving honourable mention.

The officers of this society are: President, H. Langford Warren; Vice-Presidents, A. W. Longfellow, J. Samuel Hodge, and C. Howard Walker; the Secretary and Treasurer is Mr. Frederic Allen Whiting of No. 9 Park Street, Boston.

In Providence, R. I., a centre of the great jewelry manufacturing interests of New England, there are various opportunities for the aspirant for technical proficiency in the designing and making of jewelry; there is a jewelry class in the Young Men's Christian Association, a course in the regular curriculum of the public Manual Training or Technical High School, and an important department of the Rhode Island School of Design is that devoted to jewelry designing, silversmithing, and shop work. For many years the New England Manufacturing Jewellers and Silversmiths' Association has annually offered a sum of money, to be divided into several

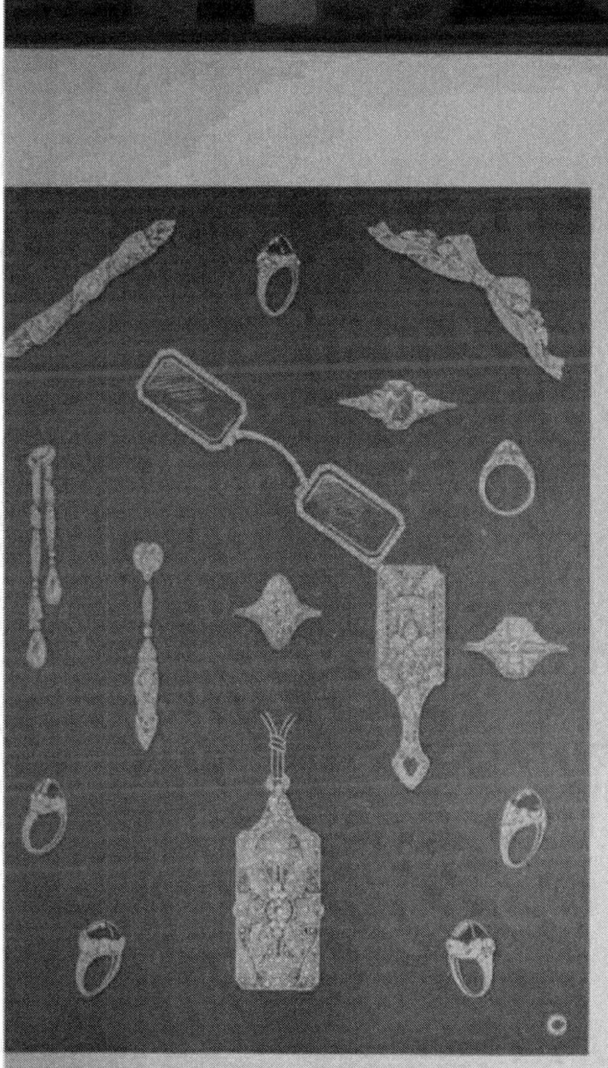

Courtesy of "Gems Creations"

STYLES OF PLATINUM DIAMOND JEWELRY, 1924

Designs by S. Kaufman

prizes, to stimulate students at this school of design to systematically study the designing of jewelry and silverware.

The Bradley Polytechnic Institute of Peoria, Ill., is an institution important in its relation to the present subject, having a jewelry course that has attained and deserves a wide reputation; the course extends over a period of from three to five months' duration. The instruction includes the making and finishing of oval and flat gold band rings, modelling for casting, signets, designing and production of jewelry, and all such repairing as is called for in ordinary jewelry store practice.

At Indianapolis, an indefatigable pioneer in the instruction of ambitious artisans in the precious metals is Mr. Charles B. Dyer, who has inaugurated a local representation of the arts and crafts movement with a school and a shop in which the hand-made jewelry of the students and graduates of the school is sold. About forty students were enrolled in the class of 1908. At a semi-annual exhibition of the students' hand-wrought products about three hundred pieces were exhibited, including bronze and copper work; the items in the exhibition were inspected with lively interest by several hundred visitors, whose commendations were enthusiastic and freely bestowed.

In response to a request, Mr. Dyer supplied an interesting account of the beginning and progress of this Middle West school, that is successfully uplifting ideals and enabling the ambitious and earnest young worker to design and make jewelry that come up to an artistic standard, as follows:

Three years ago there was formed in Indianapolis a "Society of Arts and Crafts" with a very promising membership. A house was rented and furnished and salesrooms opened. The movement grew and a large number of the right kind of people became interested. Unfortunately, however, there were so very few of the members who were craftsmen or in any way producers of salable stuff that everything had to be gotten on consignment from outside. Like so many other associations that have tried the commission plan, and through mismanagement, the society did not live long.

During its life, however, it had started a number of earnest people to thinking and had given them the desire not only to raise their standards of beauty in both useful and decorative objects, but to express their own thought and individuality. My father and I had taken great interest in the movement and had made a number of pieces of jewelry for the salesroom. When we were asked to start a class, teach the use of tools, and show how original designs could be executed in metal, we were glad to undertake the work. We started with a class of five, all of whom were art teachers in the high schools here. I might say in passing we had over seventy-five applicants this fall.

As we conduct a manufacturing jewelry business, our shop is well equipped for all kinds of metal work. We have a bench for each worker where all the small tools, hammers, wax blocks, and punches are kept and also several large vises and anvils for the large copper work. Polishers, rolls, annealing furnace, enamelling furnace, and all kinds of other tools make the shop complete enough for any work.

As the class is only a sort of pastime for us we have it at night and charge almost nothing for tuition.

The worker first designs the piece and selects the stones and material to be used. After the design has been criticised it is transferred to metal and executed. We have no class problems or lectures. All the pieces and all the criticism are individual. In that way we do not allow any worker to leave a piece until it is well executed.

Most of the workers are so interested in the work that they have their own workshops and tools at home, and a number of them have not only produced some very creditable pieces but have made good money in doing it.

At the end of each term, that is just before Christmas and in June, we have an exhibit and sale of the class work.

We send out copper plate invitations and make a social affair of it and succeed in selling most everything produced during the term. We have created a wide interest in the movement and are much encouraged to carry it along.

From many sources students are now receiving aid, encouragement, and information which but a few years ago was unheard of in America. A case in point is the offering annually by Herpers Bros., a business concern extensively engaged in the manufacture of parts of commercial jewelry, in New York City and Newark, N. J., of gold medals to the most proficient students in five leading technical schools in the United States.

At the suggestion of Hon. Oscar Straus, Secretary of the Department of Commerce and Labour, it is said: Prof. John Monaghan, for a long time a representative of the United States Government, in the consular service, has delivered series of lectures for jewellers' associations and at technical institutions which have jewelry classes or courses. While consul at Chemnitz, Germany, Prof. Monaghan devoted much time to a study of the technical schools of the German Empire.

In the opinion of Mr. Gutzon Borglum, as lately expressed in *The Craftsman*, the art school of to-day will pass and be supplanted by the school of crafts, with the predicted result that there would be immediate improvement in our wares, furniture, textiles, interior decorations, and ornaments of every kind, and that, instead of the host of unsuccessful artists of to-day, there would be successful master craftsmen, putting life and beauty into our liberal arts, invaluable citizens, and, incidentally, that these graduates of the schools of crafts would be economically independent and contented. Mr. Borglum points out that the Metropolitan Museum of Art with its collections would form a nucleus and a foundation for this useful innovation.

APPENDIX

ALPHABETICAL LIST OF GEM MINERALS

(According to Wirt Tassin)

Achirite, *see* Dioptase.
Achroite, *see* Tourmaline.
Actinolite, *see* Cat's-eye.
Adamantine spar, *see* Corundum.
Adularia, *see* Orthoclase.
Agate, *see* Quartz.
Agatized wood, *see* Quartz.
Alabaster, *see* Gypsum.
Alaska diamond, *see* Quartz.
Alexandrite, *see* Chrysoberyl.
Allanite.
Almandite, *see* Garnet.
Amazonstone, *see* Microline.
Amber.
Amethyst, *see* Quartz.
Amethyst (Oriental), *see* Corundum.
Anatase, *see* Octahedrite.
Ancona ruby, *see* Quartz.
Andalusite.
Andradite, *see* Garnet.
Anhydrite.
Apatite.
Aphrizite, *see* Tourmaline.
Apophyllite.
Asteria, *see* Corundum.
Asteria, *see* Quartz.
Aquamarine, *see* Beryl.
Aragonite, *see* Carbonate of Lime.
Arkansite, *see* Brookite.
Automolite, *see* Spinel.
Aventurine, *see* Oligoclase.
Aventurine, *see* Orthoclase.
Aventurine, *see* Quartz.
Axinite.
Azurite.
Balas ruby, *see* Spinel.
Banded agate, *see* Quartz.
Barite.
Basanite, *see* Quartz.
Beekite, *see* Quartz.

Benitoite.
Beryl.
Beryllonite.
Bloodstone, *see* Quartz.
Bone Turquoise, *see* Odontolite.
Bort, *see* Diamond.
Bottle stone, *see* Obsidian.
Bowenite, *see* Serpentine.
Brazilian diamond, *see* Quartz.
Brazilian emerald, *see* Tourmaline.
Brazilian pebble, *see* Quartz.
Bronzite.
Brookite.
Cacholong, *see* Opal.
Cairngorm, *see* Quartz.
Calcite, *see* Carbonate of Lime.
Callainite, *see* Turquoise.
Cancrinite.
Carbonado, *see* Diamond.
Carbuncle, *see* Garnet.
Carnelian, *see* Quartz.
Cassiterite.
Catlinite.
Ceylonite, *see* Spinel.
Chalcedony, *see* Quartz.
Chiastolite, *see* Andalusite.
Chlorastrolite, *see* Prehnite.
Chloromelanite, *see* Jade.
Chlorophane, *see* Fluorite.
Chlorospinel, *see* Spinel.
Chondroite.
Chromic iron.
Chrysoberyl.
Chrysocolla.
Chrysolite, *see* Olivine.
Chrysolite (Oriental), *see* Chrysoberyl.
Chrysoprase, *see* Quartz.
Cinnamon stone, *see* Garnet.
Citrine quartz, *see* Quartz.
Coal.
Cobaltite.
Cobrastone, *see* Fluorite.
Colophonite, *see* Garnet.
Congo emerald, *see* Dioptase.
Coral, *see* Carbonate of Lime.
Cornelian, *see* Quartz.
Corundum.
Crocidolite.
Cymophane, *see* Chrysoberyl.
Cyprine, *see* Vesuvianite.
Damourite.
Datolite.

DICHROISM—A LIST OF LEADING TWIN-COLOURED GEMS

Among the more important gems that display twin colours are these listed by A. H. Church in *Precious Stones* as follows:

Name of Stone.		Twin Colours.	
Sapphire	(blue),	Greenish-straw,	Blue.
Ruby	(red),	Aurora-red,	Carmine-red.
Tourmaline	(red),	Salmon,	Rose-pink.
"	(brownish-red),	Umber-Brown,	Columbine-red
"	(brown),	Orange-brown,	Greenish-yellow
"	(green),	Pistachio-green,	Bluish-green.
"	(blue),	Greenish-grey,	Indigo-blue.
Topaz	(sherry),	Straw-yellow,	Rose-pink.
Peridot	(pistachio),	Brown-yellow,	Sea-green.
Aquamarine	(sea-green),	Straw-white,	Grey-blue.
Beryl	(pale-blue),	Sea-green,	Azure.
Chrysoberyl	(yellow),	Golden-brown,	Greenish-yellow
Iolite,		Pale-buff,	Indigo-blue.
Amethyst,		Reddish-purple,	Bluish-purple.

THE MOHS TABLE OF HARDNESS

(Progressing from soft to hard.)

1.	Talc	6.	Feldspar
2.	Gypsum	7.	Quartz
3.	Calcite	8.	Topaz
4.	Fluorite	9.	Corundum
5.	Apatite	10.	Diamond

TABLE OF HARDNESS OF GEM MINERALS

(From hard to soft.)

Mineral	Hardness	Mineral	Hardness
Diamond	10	Vesuvianite	6.5
Corundum (Ruby and Sapphire)	9	Epidote	6.5
Chrysoberyl	8.5	Prehnite	6.5
Topaz	8	Pyrite	6.5
Spinel (Balas Ruby)	8–7.75	Feldspar (Amazon-stone, Moonstone, Labradorite)	6
Phenacite	7.75	Turquoise	6
Beryl (Emerald, Aquamarine)	7.75	Diopside	6
Zircon (Hyacinth)	7.5	Nephrite	5.75
Euclase	7.5	Opal	5.5–6.5
Staurolite	7.5	Moldavite	5.5
Andalusite	7.25	Obsidian	5.5
Iolite	7.25	Hematite	5.5
Tourmaline	7.25	Sphene	5.5
Garnet	7	Lapis-Lazuli	5.5
Quartz (Amethyst, Jasper, Rock Crystal)	7	Haüynite	5.5
Jadeite	6.75	Cyanite	5–7
Axinite	6.75	Dioptase	5
Chalcedony (Agate and Carnelian)	6.5	Fluorite	4
Chrysolite	6.5	Malachite	3.5
		Jet	3.5
		Amber	2.5
		Gypsum (Alabaster and Satin Spar)	2

TABLE SHOWING SPECIFIC GRAVITY OF GEM MINERALS

(*Decreasing from high to low.*)

Zircon (Hyacinth)	4.60–4.70
Almandine Garnet	4.11–4.23
Ruby	4.08
Sapphire	4.06
Cape Ruby (Garnet)	3.86
Demantoid (Garnet)	3.83
Staurolite	3.73–3.74
Pyrope (Garnet)	3.60–3.65
Chrysoberyl	3.68–3.78
Cyanite	3.60–3.70
Cinnamon Stone (Garnet)	3.60–3.65
Spinel (Balas Ruby)	3.60–3.63
Topaz	3.50–3.56
Diamond	3.50–3.52
Epidote	3.35–3.50
Vesuvianite	3.35–3.45
Sphene	3.35–3.45
Chrysolite	3.33–3.37
Jadeite	3.30
Axinite	3.29–3.30
Diopside	3.20–3.30
Dioptase	3.29
Andalusite	3.17–3.19
Apatite	3.16–3.22
Hiddenite	3.15–3.20
Green and Blue Tourmaline	3.11–3.16
Euclase	3.05
Fluospar	3.02–3.19
Nephrite	3.00
Phenacite	2.98–3.00
Red and Colourless Tourmaline	2.94–3.08
Turquoise	2.60–2.80
Labradorite	2.70
Beryl	2.68–2.75
Emerald	2.67

Rock Crystal	2.65–2.66	Obsidian	2.50–2.60
Smoky Quartz	2.65–2.66	Moonstone (Adularia)	2.55
Amethyst	2.65–2.66	Lapis-lazuli	2.40
Jasper	2.65–2.66	Moldavite	2.36
Chrysoprase	2.65–2.66	Opal	2.19–2.20
Iolite	2.60–2.65	Jet	1.35
Chalcedony	2.60	Amber	1.00–1.11
Agate	2.60		

REFRACTION

The refractive indices of the more important precious and semi-precious stones are given in the following table, the values for singly refracting stones being indicated by *n*, and the greatest and least values for doubly refracting stones by *n y* and *n z*, respectively. In both cases the values apply to the middle rays of the spectrum. The strength of the double refraction of each stone is indicated by *d-n y—n z*, that is, by the difference between the greatest and the least refractive indices of the stone.

SINGLY REFRACTING PRECIOUS AND SEMI-PRECIOUS STONES

	n		*n*
Diamond	2.43	Spinel	1.72
Pyrope	1.79	Opal	1.48
Almandine	1.77	Fluor-spar	1.44
Hessonite	1.74		

DOUBLY REFRACTING PRECIOUS AND SEMI-PRECIOUS STONES

	n y	*n z*	*d*
Zircon	1.97	1.92	0.05
Ruby, Sapphire	1.77	1.76	0.01
Chrysoberyl	1.76	1.75	0.01
Chrysolite	1.70	1.66	0.04
Tourmaline	1.64	1.62	0.02
Topaz	1.63	1.62	0.01
Beryl	1.58	1.57	0.01
Quartz	1.55	1.54	0.01

TRANSPARENCY OF GEMS UNDER RÖNTGEN (X) RAYS

Completely transparent
- Amber
- Jet
- Diamond

Strongly transparent
- Corundum

Transparent
- Opal
- Andalusite
- Cyamite
- Chrysoberyl

Semi-transparent
- Quartz
- Labradorite
- Adularia
- Topaz

Slightly transparent
- Spinel
- Essonite (Garnet)
- Fluorite

Almost opaque
- Gypsum
- Turquoise
- Tourmaline
- Calcite

Opaque
- Almandite (Garnet)
- Beryl
- Epidote
- Rutile
- Hematite
- Pyrite
- Zircon

A CARAT'S WEIGHT IN VARIOUS LOCALITIES

The weight of a carat is rated differently in various localities where the diamond industry is important. On an average, the carat does not differ in value much from the fifth of a gram of the metric system (200 milligrams), or about three and one sixth English grains.

The fractions of the carat used in weighing precious stones are ½, ¼, ⅛, and so on down to one sixty-fourth; this fraction of a carat of 205 milligrams is equal to 3.203 milligrams. The fourth part of a carat is known as a grain; not a Troy weight grain, however, but a "pearl grain"; although this is rarely used as a unit. In France 144 carats equal one ounce. Efforts are continually being made to reconcile these variations of weight in the use of the term "carat," and also to substitute the gram of the metric system for the carat, and it is hoped that eventually the weighing of precious stones may be universally standardised.

The exact values in milligrams of the carat at different places are tabulated as follows:

Locality	Milligrams
Amboina	197.000
Florence	197.200
New York	205.000
Batavia	205.000
Borneo	205.000
Leipzig	205.000
Spain	205.393
London	205.409
Berlin	205.440
Paris	205.500
Amsterdam	205.700
Antwerp	205.300
Lisbon	205.750
Frankfurt-am-Main	205.770
Venice	207.000
Vienna	206.130
Madras	207.353
Livorno	215.990

An International Committee of Weights and Measures has finally adopted the recommendations of the various associations of jewellers and diamond merchants, and has officially sanctioned a fixed uniform weight value of 200 milligrams for the carat.

The carat, as a weight (it should be remembered that the word Karat is also used in the jewelry trade to denote the fineness of gold), is used for weighing the precious stones. The weights of some semi-precious stones in the trade are reckoned in pennyweights. Pearls are weighed and their values calculated by the grain (¼ carat). Diamonds are designated in the trade as "grainers," "two grainers," etc.; a "four grainer" is a diamond weighing one carat. The practice in the United States has been to calculate the carat at 205 milligrams or 3.210 grains Troy. The new metric carat, which will probably eventually become the universally recognised standard, will weigh 200 milligrams or 3.130 grains.

CRYSTALLOGRAPHY

SYSTEMS OF CRYSTALLINE FORM

1. The Cubic System	with	9	planes	of symmetry
2. The Hexagonal System	"	7	"	"
3. The Tetragonal System	"	5	"	"
4. The Rhombic System	"	3	"	"
5. The Monoclinic System	"	3	"	"
5. The Monoclinic System	"	1	"	"
6. The Triclinic System	"	0	"	"

"BIRTH-STONES"

A RHYMING LIST OF NATAL GEMS POPULARLY IDENTIFIED WITH THE MONTHS

(*Substantially as published by* **Wirt Tassin** *and other authorities*)

JANUARY

By those in January born
No gem save *garnet* should be worn;

Should wear *topaz* of amber hue,
Emblem of friends and lovers true.

DECEMBER

If cold December gave you birth,
The month of snow and ice and mirth,
Place on your hand a *turquoise* blue—
Success will bless whate'er you do.

BIBLIOGRAPHY

BIBLIOGRAPHICAL NOTE

SINCE one book cannot possibly comprehend all the phases of a large subject, it may be of service to some of our readers to supply a full bibliography, like that which follows, a bibliography that will enable them easily to acquire information on special phases, or advance to a liberal education on the entire subject. It should be said, however, that an absorption and assimilation of all that was ever printed about gems, even with the aid of illustrations—line, half-tone, and colour-work of the most advanced stage of reproductive pictorial art—cannot thoroughly inform the student without close study of the gem stones and cut gems themselves.

The most comprehensive book about gems ever written is undoubtedly *Precious Stones* by Dr. Max Bauer. The original of this monumental work was first published in parts under the title *Edelsteinkunde* in 1895 and 1896, in Germany, but was subsequently translated into English by L. J. Spencer, of the mineral department of the British Museum, and published in 1904 in London, and a little later in this country. With interest, pride, and pleasure Americans may read the initial sentence of Dr.

Bauer's introduction to his book, as follows: "The desire of the publishers to present to the German public a work on precious stones, similar in character to that admirably supplied in American literature by George Frederic Kunz's *Gems and Precious Stones of North America*, gave the initiative to the writing of the present book." That the foremost expert on American gems should be an American, designated as its official authority by the United States Government, and accepted as such abroad, and that this American should possess the literary ability to disseminate the knowledge he has gathered in a popular as well as strictly scientific fashion, and should have directly caused the production of the most authoritative book on the gem subject, may be a source of satisfaction to his compatriots who are patriotic in all things as well as admirers of gems.

The basis of much of the information extant about gems is the old, but reliable and still standard, *A System of Mineralogy*, by James Dwight Dana, published in 1837, in New Haven, Conn. This text-book, supplemented with Bauer's great book, and with the addition of Kunz's *Gems and Precious Stones of North America* to cover the phase of the general subject involving American gems, contains all important facts about gems and gem minerals, exclusive of recent mineralogical and other pertinent scientific discoveries. A valuable associate to this trio would be the *Descriptive Catalogue of the Collections of Gems in the United States National Museum*, by Wirt Tassin, Assistant Curator of the Division of Mineralogy. This was reprinted by the Government Printing Office at

Washington, in 1902, from *The Report of the United States National Museum for 1900*. This report is out of print as a separate publication, but would be available through the acquisition of the annual report named, or should be obtainable in any extensive public library.

The following bibliography combines two lists of works on the subject in hand compiled respectively by Mr. A. P. Griffin, Chief Bibliographer of the Division of Bibliography, Library of Congress, and Mr. Wirt Tassin, to both of whom the author gratefully acknowledges his indebtedness.

BIBLIOGRAPHY

ABDALAZIZ (AHMED BEN). *Treatise on jewels.*

ABEN EZRA (RABBI). *Commentarium in Decalogum.* Basel (Basle), 1527.

ABICH (H.). *De Spinello.* Berolini (Berlin), 1831.

ADLER (C., and CASANOWICZ). *Precious stones of the Bible.* [In Biblical Antiquities, Report, U. S. National Museum, 1896, p. 943.]

AGOSTINI (L.). *Gemmæ et sculpturæ antiquæ.* Franequeræ (Franecker), 1699.

AGRICOLA (G.). *De ortu et causis subterraneorum de natura eorum quæ effluunt ex Terra.* Basel (Basle), 1558.

AGRIPPA (H. C.). *Philosophie occulte.* [Translated by Levasseur.] La Haye (The Hague), 1655. Contains material relating to the mystical properties of gems.

ALAMUS AB INSULIS (ALAIN DE LISLE). *Dicta alani, etc.* Lugduni-Batavorum (Leyden), 1599. An alchemical treatise containing material relating to the mystical properties of gems. A lamus ab Insulis, b. 1114, d. 1202, was the earliest Flemish alchemist.

ALBERTUS MAGNUS. *Die mineralibus.* [In his opera, v. ii.] Lugduni (Leyden), 1651.

ALBERTUS (*Cont.*). *De Vertutibus herbarum, lapidum animalum, etc.* Various editions.

——. *Les admirable secrets d' Albert le grand, etc.* Lyon, 1758. Contains extracts from the works of Albertus Magnus, relating to the magical and medicinal properties of gems.

ALCOT (T.). *Gems, talismans, and guardians.* New York, 1886.

ANDRADA (M. D'). *An account of the diamonds of Brazil.* [In Nicholson's Journal, i., 1797, 24.]

ANTIDOTARIO DE FRA D. D'E. Napoli (Naples), 1639. A treatise on pharmacy, containing a few accounts of the virtues of gems.

ARGENVILLE. *Traité de l'Oryctologie.* Paris, 1740.

ARGENVILLE (A. J. D. D'). *De l'Histoire Naturelle éclaircie dans deux de ses parties principales: la Lithologie et la Conchologie.* Paris, 1742.

ARISTOTLE. His works, especially the *Meteorology* and *Wonderful things heard of.* Aristotle was born about 384 B.C., and died about 322 B.C.

——. *Lapidarius.* [De novo Græco translatus, Lucas Brandis.] Regia Mersbourg (Merseburg), 1473.

ARNALDUS DE VILLANOVA. *Chymische schriften, etc.* [Translated by Johannem Hippdamum.] Wien (Vienna), 1742. See also: Hermetischer Rosenkranz, Pretiosa Margarita, Manget, Theatrum Chemicum, etc. The several writings of this alchemist (also called Villanovanus, Arnald Bachuone, A. de Villeneuve, and Arnaldus Novicomensis) contain much concerning the occult, medicinal, and other properties of gems.

ARNOBIO (CLEANDRE). *De Tesoro delle Gioie, trattato maraviglioso.* Venit. (Venice), 1602.

ATHANAEUS. *Deiphriosophistæ* [*Banquet des Philosophes*], translated by Dalechamp. Paris, 1873.

AUBREY (J.). *Miscellanies.* London, 1857. Contains an account of the use of the beryl in divination.

AVICENNA (ABOU-ALI-ALHUSSEIN-BEN-ADLOULAH). *Canonnes Medicinæ* [Lat. reddit.] Ven. (Venice), 1843. Contains material relating to the medicinal and magical virtues of gems.

BABELON (ERNEST). *La gravure en pierres fines camées et intailles.* Paris: Quentin [1894]. 320 pp. Illustrations. 8°. (Bibliothèque de l'enseignement des beaux-arts.)

——. *Histoire de la gravure sur gemmes en France depuis les origines jusqu' à l'époque contemporaine; ouvrage illustré de gravures dans le texte et accompagné de XXII planches en phototypie.* Paris; Société de propagation des livres d'art, 1902. [iii]–xx., 262 pp. (2), Illustrations. XXII plates. 4°.

BABINGTON (CHARLES). *A systematic arrangement of minerals, their chemical, physical, and external characters.* London, 1795.

BACCI (ANDREA). *Le XII Piètre preziose.* Roma (Rome), 1587.

——. *De Gemmis et Lapidibus pretiosis, tractatus ex Ital.* Lingua Lat. red. Francof. (Frankfurt), 1605.

——. *De Gemmis ac Lapidibus pretiosis in S. Scriptura.* Roma (Rome), 1577; 8°, Franc. (Frankfurt), 1628.

BACON (ROGER). *Opera Quædam hactenus inedita.* [Edited by J. S. Brewer.] London, 1859. The appendix —Epistola . . . de secretis operibus artis et naturæ— contains some material relating to the magical and alchemical virtues of certain gems.

BALL (V.). *On the occurrence of diamonds in India.* [In Geology of India, 3 vols., pp. 1–50, 1881.]

——. *On the mode of occurrence and distribution of diamonds in India.* [In Proc. R. Dublin Soc., ii., p. 551; also Jour. R. Geol. Soc. Ireland, vi., p. 10.]

——. *On the geology of the Mahanadi basin and its vicinity.* [In Records of the Geological Survey of India, x., p. 167: map.]

——. *A manual of the geology of India.* Calcutta, 1881.

——. *On the identification of certain diamond mines in India which were known and worked by the ancients, especially those visited by Tavernier. With a note on the history of the Koh-i-nur.* [In Journal of the Asiatic Society of Bengal, l., 1881, p. 31; Report British Association for 1882, p. 625; and Nature, xxiii., p. 490, 1882.]

BALL (*Cont.*). *On the diamond, etc., of the Sambâlpúr district.* [In Records of the Geological Survey of India, x., p. 186: map.]

BAPST (G.). *Les joyaux de la couronne.* [In Revue des Deux Mondes, 1886, p. 861.]

BARBET (Ch.). *Traité complet des pierres précieuses.* Paris, 1858.

BARBET DE JOUY (HENRI). *Les gemmes et joyaux de la couronne au Musée du Louvre;* expliqués par M. Barbet de Jouy . . . dessinés et gravés à l'eau-forte d'après les originaux par J. Jacquemart . . . introduction par A. Darcel. Paris: L. Techener 1886. (12) pp., 60 plates. F°.

BARRERA (Mme. DE). *Gems and jewels.* London, 1860.

BAUER (MAX). *Edelsteinkunde.* Leipzig, 1896.

——. *Precious stones.* Translated from the German of the above with additions by L. J. Spencer. London: C. Griffin and Company, 1904, xv., (1), 627 pp. Illustrations. Plates (partly colored). 8°. Philadelphia, J. B. Lippincott & Co., 1904.

BAUMER (J. W.). *Historia Naturalias Lapidum pretiosorum omnium, etc.* Franc. (Frankfurt), 1771.

——. *Naturgeschichte aller Edelstein, wie auch der Erde und Steine, so bisher zur artznei sind gebraucht worden.* Aus dem Latein von Karl, Freih. von Meidinger. Wien (Vienna), 1774.

BAUMHAUER (E. H. VON). *Diamonds.* [In Ann. Phys. Chem., 2 ser., i., 1877, p. 462.]

BEARD (C. P.). *Traité des pierres précieuses.* Paris, 1808.

BECHAI BEN ASCHAR. *Biur al Hattorah—Exposition of the Law of Moses, a commentary on Exodus* xxviii., 17–20. A.M. 5207 (A.D. 1447). Contains an account of the virtues and properties of gems.

BECHER (JOHANN JOACHIM). *Physica Subterranea.* Lipsiæ (Leipzig), 1739. An alchemical work.

BECK (R.). *Die diamantenlager stätte von Newland in Griqualand West.* [In Zeits. für Prakt. Geol., 1898, p. 158.]

BEHRENS (TH. H.). *Sur la cristallisation du diamant.* [In Arch. Neerl., xvi., p. 376, 1881.]

BEKKERHEIM (KARL). *Krystallographie des Mineralreichs.* Wien (Vienna), 1793.

BELLEAU (RENÉ). *Les amours et nouveaux échanges des pierres précieuses.* Paris, 1576.

BELLERMAN (J. J.). *Die Urim und Thummin.* Berlin, 1824.

BENIAM (MUTAPHIA). *Sententiis sacro medicis.* Hamburg, 1640. Contains material relating to the astrological virtues of gems.

BERQUEN (ROBERT DE). *Les Merveilles des Indes Orientales et Occidentales, ou nouveau. Traité des Pierres précieuses, et des Perles.* Paris, 1661.

BESONDERE. *Geheimnisse eines wahren Adepti von der Alchymie, etc.* Dresden, 1757. An alchemical treatise.

BEUMEMBERGER (J. G.). *Der Volkomene Juwelier.* Weimar, 1828.

BIELHE (VON). *Ueber die Bernstein-Gräbereien in Hinter Pommern.* Berlin, 1802.

BILLING (A). *Science of gems, coins, and medals.* New York, 1875.

BIRDWOOD (G. C. M.). *Industrial arts of India.* Vol. ii., pp. 17–32, 1881.

BISHOP (HEBER REGINALD). *The Bishop collection; investigations and studies in jade.* New York, 1900. (6), 378 pp. 8°. Bibliography, pp. 367–370.

BLEASDALE (J. J.). *Gems and precious stones found in Victoria.* [In an essay in Official Record, Inter-Colonial Exhibition, Melbourne, 1867.]

BLUM (J. R.). *Verzeichniss der geschnitten Steine in dem Königl. Museum zu Berlin.* Berlin, 1827.

——. *Lithurgik, oder mineralien und Felsarten nach ihrer Anwendung in Oekon., artist, und Technischer Hinsicht systematische abgehandelt.* Stutgart, 1840.

BLUM (R.). *Die Schmucksteine.* Heidelberg, 1828.

——. *Taschenbuch der Edelsteinkunde.* Stutgart, 1840.

BLUMENBERG. *Dissertatio Medica de Succino.* Jena, 1682.

BLUMHOF (J. C.). *Lehrbuch der Lithurgik.* Frankfurt, 1822.

BOETIUS (ANSELMUS). *Tractatus de Lapidibus.*

BOLNEST (E.). *Aurora chymica, or a rational way of preparing animals, vegetables, and minerals for a physical use; by which preparations they are made most efficacious, safe, pleasant medicines for the preservation and restoration of the life of man.* London, 1672.

BONDARY (JEAN DE LA TAILLE DE). *Blason des Pierres précieuses.*

Booke of the Thinges that are brought from the West Indies. [English translation, 1580.] 1574. Contains an account of the virtues of the bloodstone.

BOOT (ANSELMUS BOETIUS DE). *Le parfaict joaillier, ou histoire des Pierres, de nouveau enrichi de belles annotations par André Toll.* [Translated from Latin by J. Bachou] Lyon, 1644.

——. *Gemmarum et Lapidum Historia.* Hanover, 1690.

BOOT (B. DE). *Lap. Gemmarum et Lapidum Historia.* Jena, 1647. The first edition published at Jena in 1609; the second enlarged by A. Toll, Lugduni Bat. [Leyden], 1636, contains much concerning the mystical and medicinal properties of gems.

BORDEAUX (A.). *Les mines de l'Afrique du Sud.* Paris, 1898.

BORN (BARON INIGO). *Schneckensteine, oder die Sächsischen Topasfelsen.* Prag, 1776.

BOURNON (COMTE DE). *An analytical description of the crystalline forms of corundum from the East Indies and China.* [In Phil. Trans.: Abr., xviii., p. 368, 1798.]

——. *Description of the corundum stone, and its varieties, commonly known as oriental ruby, sapphire, etc.* [In Phil. Trans., 1801, p. 223.]

——. *A descriptive catalogue of diamonds in the cabinet of Sir Abraham Hume.* London, 1815.

BOUTAN (M. E.). *Diamant.* [In Frémy's Encyclopédie Chimique.]

——. *Le Diamant.* Paris, 1886. Contains a very full bibliography.

BOYLE (ROBERT). *Experiments and considerations upon*

color, with considerations on a diamond that shines in the dark. London, 1663.

——. *Essay about the origin and virtues of gems.* [In his works, v. iii., 1772.]

——. *Exercitatio de origine et viribus gemmarum.* London, 1673.

——. *An essay about the origin and virtues of gems, with some conjectures about the consistence of the matter of precious stones.* London, 1672. [Another edition in 1673.]

BRARD (C. P.). *Traité des Pierres Précieuses, des Porphyres, Granits et autres Roches propres à recevoir le poli.* 1808.

——. *Minéralogie appliquée aux arts.* Paris, 1821.

BRITISH MUSEUM. *Catalogue of Gems in the British Museum (Department of Greek and Roman Antiquities).* 1888.

BRITTEN (EMMA H.). *Art Magic; or mundane, submundane, and supermundane spiritism.* Contains accounts of mystical properties of gems.

BRONGNIART (ALEXANDRE). *Traité de minéralogie, avec application aux arts.* Paris, 1807.

BROWN (C. B. and J. W. JUDD). *The rubies of Burma.* [In Phil. Trans. Roy. Soc. London, clxxxvii., p. 151–228.] A very elaborate and complete account of the physical features, geology, and geographical distribution of the ruby-bearing rocks of the district.

BRUCKMANN (U. F. B.). *Abhandlung von Edelsteinen.* Braunschweig (Brunswick), 1757–73.

——. *A treatise on precious stones.* 1775.

——. *Gesammelte und eigene Beiträge zu seiner Abhandlung von Edelsteinen.* Braunschweig (Brunswick), 1778.

BUCHOZ (——). *Les Dons merveilleux et diversement coloriés de la Nature dans le Règne Minéral.* Paris, 1782.

BUFFUM (W. A.). *The tears of the Heliades or amber as a gem.* New York, 1900.

BURCH (A.). *Handbuch für Juweliere.* Weimar, 1834.

BURNHAM (S. M.). *Precious stones.* Boston, 1886.

BURTON (R. F.). *Gold and diamond mines.* [In his Explorations of the Highlands of Brazil, 1869.]

BUTLER (G. MONTAGUE). *A Pocket Handbook of Minerals.* John Wiley & Sons, New York.

CADET (LE JEUNE). *Mémorie sur les Jaspes et autres Pierres Précieuses de l'ile de Corse.* Bastia, 1785.

CAESALPINUS (ANDREAS). *De metallicis Libri tres.* Rom. (Rome), 1496.

CAHAGNET (L. A.). *Magie magnetique.* Paris, 1838. A spiritualistic work containing material relating to the occult properties of gems.

CAIRE (A.). *La Science des pierres précieuses appliquée aux arts.* Paris, 1833.

CAPELLER (MAUR. ANT.). *Prodomus crystallographiæ, de crystallis improprie sic dictis commentarium.* Lucernæ (Lucerne), 1723.

CARDANUS (HIERONYMUS). *De Lapidibus preciosis;* also *De Subtilitate.* These contain accounts of the magical and medicinal properties of gems.

CAROSI (JOHANN). *Sur la Generation du Silex et du Quartz.* Cracov, 1783.

CARTON (J.). *Englischen Juwelier, Kenntniss, Werthund Preisschatzung aller Edelsteine, Perlen, Corallen, ins Deut. ubersetzt nach der* 10 *ed.* Gratz, 1818.

CASTELLANI (A.). *Gems, notes, and extracts.* [Translated from the Italian, by Mrs. J. Brogden.] London, 1871.

Catalog des Bijoux nationaux. Paris, 1791.

CATTELLE (W. R.). *The Pearl.* J. B. Lippincott Co., Philadelphia, 1907.

CELLINI (BENEVENUTO). *Trattato del' Oreficeria.*

——. *Del Arte del Gioiellare.* Fior. (Florence), 1568.

CHAFER (——). *Note sur la région diamantifère de l'Afrique Australe.* Paris, 1880.

——. *On the occurrence of diamonds in India.* [Comptes Rendus, 1884, p. 113.]

CHAND (GULAL). *Essay on diamonds.* Lucknow, 1881.

Characteristics and Localities of the Principal Precious Stones. Supplement to The Jeweller's Circular: The Jeweller's Circular Publishing Co., New York.

CHURCH (A. H.). *Precious and curious stones.* [In Spectator, July 9, 1870.]
——. *Townsend Collection.* [In Quart. Jour. Science, Jan., 1871.]
——. *Precious stones.* London, 1882.
——. *Discrimination, etc., of precious stones.* [In Jour. Soc. Arts., xxix., p. 439.]
——. *Physical properties of precious stones.* [In Proc. Geol. Assoc., v., No. 7.]
——. *Colours of precious stones.* [In Magazine of Art., i., p. 33.]
CLAREMONT (LEOPOLD). *Precious stones.* Philadelphia and London. J. B. Lippincott Company, 1903. 224 pp., xix. plates. 8°.
——. *The gem-cutter's craft.* London. G. Bell and Sons, 1906. xv., (1), 296 pp. Illustrations. Plates. Tables. 8°.
CLAUDER (G). *Schediasma de tinctura universali, vulgo lapis philosophorum dicta, etc.* Norimbergæ (Nuremberg), 1736. An alchemical treatise containing 13 folding tables having a list of minerals with their properties grouped under the following heads: Nomen, Substantia, Color, Pondus, Natura, Præparatio, Tractatio, Contenta.
CLAVE (ESTIENNE). *Paradoxes, ou Traittez Philosophiques des Pierres et Pierreries, contre l'opinion volgaire.* Paris, 1635.
CLUTIUS (AUGERIUS). *Calsvee, sive Dissertatio Lapidis Nephritici, seu jaspidis viridis, naturam, proprietates, et operationes exhibens Belgice.* [Amsterdam, 1621, et Lat. per Gul. Lauremberg, fil.] Rostochii (Rostock), 1627.
COHEN (E.). *Ueber Capdiamanten.* [In Neues Jahrbuch, i., p. 184, 1881.]
COHEN (M.). *Beschreibendes Verzeichniss einer Sammlung von Diamanten.* Wien (Vienna), 1822.
COLLINI (COSMUS). *Journal d'un Voyage, qui contient différentes observations minéralogiques, particulierment sur les agates, avec un détail sur la menière de travailler les agates.* Mannheim, 1776.

COLONNE (FRANÇOISE MARIA POMPÉE). *Histoire Naturelle de l'Univers.* [4 vols.] Paris, 1734.

CORSI (FAUST). *Delle Piedre antiche libri quattro.* Roma (Rome), 1828.

CROLY (G.) *Gems; etched by R. Dagley, with illust., in verse.* London, 1822.

CROOKES (SIR W.). *Diamonds.* [In Proc. Roy. Inst., 1897, p. 477.]

CROOKES (WILLIAM). *Diamonds.* [In Report Smithsonian Institution, 1897, p. 219.]

——. *On radiant matter.* [In Chemical News, xl., pp. 93, 104, and 127.] Contains results of experiments on the phosphorescence of the diamond, ruby, and other minerals.

Curiose speculationen. Leipzig, 1707.

CURL (MARTHA A.). *Ancient gems.* [In American Antiquarian, xxii., p. 284, 1900.]

DALL (W. H.). *Pearls and pearl fisheries.* [In American Naturalist, 1883, pp. 579, 731.]

DANA (E. S.). *On the emerald green spodumene (Hiddenite) from Alexander County, North Carolina.* [In Am. Jour. Science, 1881, xxii., p. 179.]

DANA (E. S., and H. L. WELLS). *Description of the new mineral, beryllonite.* [In Am. Jour. Science, 1889, xxxvii., p. 23.]

DAUBRÉE (M.). *Rapport sur un mémoire de M. Stanislas meunier ayant pour titre: Composition et origine du sable diamantifère de Du Toits Pan.* [In Comptes Rendus, lxxxiv., p. 1124.] A summary of the subject to the date.

DAVENPORT (CYRIL JAMES H.). *Cameos.* London: Seeley & Co. New York: The Macmillan Co., 1900, viii., 66 pp. Frontispiece. Illustrations. Plates partly coloured. Portrait. 4°. (The Portfolio; monographs, No. 41.)

DAVY (HUMPHRY). *Some experiments on the combustion of the diamond and other carbonaceous substances.* [In Phil. Trans., 1814, p. 557.]

De Lapidibus, Avibus et Arboribus Indiæ, Arabiæ, et Africæ. [Harleian manuscripts.]

DERBY (O. A.). *The geology of the diamantiferous region of the province of Paraná, Brazil.* [In Proc. Am. Phil. Soc., xviii., p. 251; also Am. Jour. Science, 1879, xviii., p. 310.]

——. *On the occurrence of diamonds in Brazil.* [In Am. Jour. Science, 1882, xxiv., p. 34.]

——. *Notes on certain schists of the gold and diamond region of eastern Minas Geraes, Brazil.* [In Am. Jour. Science, 1900, x., p. 207.]

Diamantengrabereien in Südafrika. [In Zeits. deutsch. Ing. Arch. Ver., xxvi., 1883, p. 565.]

Diamond, Description of the. [In Phila. Trans, Abr., ii., 1708, p. 405.]

Diamond, The, or the pest of a day. London, 1797.

Diamond, The artificial production of. [In Nature, xxii., 1880, pp. 404, 421.]

Diamond (The). [In Westminster Review, Jan., 1883.]

Diamond, Fresh . . . discoveries in New South Wales. [In Iron, xxiii., p. 249, 1884.]

Diamond. Papers and notes on the genesis and matrix of the . . . by the late Henry Carvill Lewis, edited by H. C. Bonney. London, 1897.

Diamonds. [In Nature, Aug. 5, 1887, p. 325.]

Diamond Cutting. [In 13 Annual Report of the U. S. Commissioner of Labor.] Deals with subjects relating to the comparison of hand and machine work.

Diamond mining at Kimberley, South Africa. [In Geol. Mag., x., 1883, p. 460.]

DINGLEY (ROBERT). *On gems and precious stones, particularly such as the ancients used to engrave on.* [In Phil. Trans.: Abr., ix., 1747, pp. 345.]

DIEULAFAIT (L.). *Diamants et Pierres Précieuses.* Paris, 1871.

——. *Diamonds and precious stones; a popular account of gems.* New York, 1874.

Dioscorides materia medica. Written about A.D. 50. A portion of the work treats especially of the medicinal properties of minerals.

DIXON (A. C.). *Rocks and minerals of Ceylon.* [In

Jour. Ceylon Branch Roy. Asiatic Soc., vi., No. 22, p. 39. Colombo.]

DOELTER (C.). *Edelstein Kunde. Bestimmung und Untersuchung der Edelsteine und Schmucksteine Kuenstliche Darstellung der Edelsteine.* Leipzig, 1893.

DOLCE (LUDOVICO). *Libré tre, nei Quali si tratta delle diverse sorti delle gemme che produce la Natura.* Ven. (Venice), 1564.

DÖLL (E.). *Zum vorkommen des Diamants in Itakolumite Brasiliens und in den Kopjen afrikas.* [In Verh. k.-k. geol. Reichs., 1880, p. 78.]

DRÉE (——). *Catalogue de Musée Mineralogique.* Paris, 1811.

——. *Voyage aux mines de diamants dans le Sud de l'Afrique.* [In Tour du Mond, Nos. 931–933, 1878.]

DU CHESNE (J.). *A Briefe Aunswere of Iosephus Quercetanus Armeniacus, etc.* London, 1591. Contains a second part "concerning the use of mineral medicines."

DU MERSAN (T. M.). *Histoire du cabinet des medailles, Pierres Gravées, etc.* Paris, 1838.

DUMONT (and JOURDAN). *Pierres précieuses.*

DUNN (E. J.). *Notes on the diamond fields of South Africa.* [In Quart. Jour. Geol. Soc., xxxiii., p. 879, and v. 37, p. 609.]

DUTENS (LEWIS). *Des pierres précieuses et des pierres fines, avec les moyens de les connoitre et de valuer.* [In his Œuvres, ii.] Londres, 1776.

ECCHELLENSIS (ABRAHAM). *Versio Durrhamani de medicis Virtutibus animalum, plantarum et gemmarum.* Paris, 1647.

ECKERMAN (N). *Electra, oder die Enstehung des Bernsteins.* Halle, 1807.

EEKHEL (J. H.). *Choix des Pierres gravées du Cabinet Impérial des Antiques.* Vienne (Vienna), 1788.

EICHORN (J. G.). *Die gemmis scalptis Hebraeorum.* [In Goettingen Ges. d. Wiss. Comm., 1811–13.]

EKEBERG (ANDREW GUSTAVUS). *Dissertatio de Topazio.* Upsal (Upsala), 1796.

ELLIOTT (JOHN). *On the specific gravity of diamonds.* [In Phil. Trans.: Abr., ix., 1745, pp. 147.]

EMANUEL (H.). *Diamonds and precious stones.* London, 1865. Contains a very full bibliography.

ENCELIUS (CHRISTOPH). *De Re Metallica, hoc est, de origine varietate et natura corporum metallicorum, Lapidum, Gemmarum atque aliarum quae ex fodinis eruuntur, Libri III.* Francf. (Frankfurt), 1551.

ENGELHARDT (AR. VON). *Die Lagerstatte der Diamenten im Ural-Gebirge.* Riga, 1830.

EPIPHANIUS. *De duodecim Gemmis in veste Aaronis.* [Gr. Lat. cum corollario Gesneri.] Tig. (Turin), 1565.

ERCKER (L.). *Aula Subterranea.* 1595.

ERMANN. *Beitrage zur Monographie des Marekasit, Turmalin, und Brasilianischen Topas.* Berlin, 1829.

FABRE (P. J.). *L'Abrégé des secrets chymiques, ou l'on void la nature des animaux, végétaux, et minéraux entiérement découverte.* Paris, 1636.

FALLOPIUS (G.). *De Medicatis Aquis atque de Fossilibus, tractatus ab Andrea Marcolini collectus.* Venetia (Venice), 1564.

FARRINGTON (OLIVER CUMMINGS). *Gems and gem minerals.* Chicago, A. W. Mumford, 1903. xii., 229 pp. Illustrations. Plates. Maps. 4°.

FERGUSON (A. M. and J.). *All about gold, gems, and pearls in Ceylon and southern India.* London, 1888.

FERNEL (JOHN FRANCIS). *Pharmacia, cum Guliel, Plantii et Franc. Saguyerii Scholiis.* Hanov. (Hanover), 1605.

FEUCHTWANGER (L.). *Treatise on gems in reference to their practical and scientific value.* New York, 1838.

——. *Popular treatise on gems in reference to their scientific value: a guide for the teacher, etc.* New York, 1859.

FICORINI (F.). *Gemmæ antiquæ adnot. N. Galeotti.* Romæ (Rome), 1757.

FINOT (L.). *Les Lapidaires Indiens.* Paris, 1896. Contains eight different Sanskrit books of the art of the Indian lapidary, two of which are translated. The gems are described with reference as to origin, their value as charms, and also as to their occurrence, colour, class, and value.

FISCHER (G. DE WALDHEIM). *Essai sur la Turquoise et sur la calaite.* Moscou (Moscow), 1810.

——. *Essai sur la Pellegrina, ou la Perle incomparable des frères Zozima.* Moscou (Moscow), 1818.

FLADE (C. G.). *De re metallica Midianitarum et Phœnicornum.* Lipsiæ (Leipzig), 1806.

FLADUNG (J. A. F.). *Versuch über die Kennzeichen der Edelsteine und deren vortheilhaftesten Schnitt.* Pesth (Budapest), 1819.

——. *Edelsteinkunde.* Wien (Vienna), 1828.

FONTENAY (——). *Bijoux anciens et modernes.*

FONTENELLE (——). *Nouveau manuel complet du bijoutier.* Paris, 1855.

FORSTER (J. A.). *Diamonds and their history.* [In Jour. Microscopy Nat. Science, iii., 1884, p. 15.]

FOWLE (——). *Occurrence of diamonds in China.* [In U. S. Consular Report, No. 198, 1897, p. 384.]

FOUQUÉ (F. and M. LÉVY). *Synthèse des minéraux.* Paris, 1871.

FRÉMY (E. and TEIL). *Artificial production of precious stones.* [In Jour. Soc. Arts, xxvi., 1878.]

——. *Sur la production artificielle du corindon du rubis et de différents silicates cristallisés.* [In Comptes Rendus, lxxxv., p. 1029.]

FRIEDLÄNDER (I.). *Artificial production of diamond in silcates corresponding to the actual mode of occurrence in South Africa.* [In Geol. Mag., p. 226, 1898.]

FRISCHOLZ (J.). *Lehrbuch der Steinschneidekunst, fur Steinschneider, Graveurs, etc., und Jedens, welcher sich über die Veredlung der Steine zu unterrichten wünscht.* München (Munich), 1820.

FURTWÄNGLER (ADOLPH). *Die antiken Gemmen. Geschichte der Steinschneidekunst im klassischen Altertum.* Leipzig, Berlin: Giesecke & Devrient, 1900. 3 vols. Illustrations. Plates. F°.

GALAMAZAR (——). *Liber de Virtutibus lapidum pretiosorum quem scripsit Galamazar, Thesaurarius Regis*

Babylonie, ipso presenti et precipiente. [In Harleian Manuscripts.]

GAUTIER (J.). *Untersuchung über die Entstehung, Bildung und den Bau des Chalcedons, etc.* Jena, 1809.

Gems. [In Spon's Encyclopedia of the Industrial Arts, p. 1042.]

GERHARD (C. A.). *Disquisitio physico-chemica granatorum Silesiæ atque Bohemiæ.* [Inaug. Diss. Frankfurt a. d. Oder, 1760.]

GESNER (CONRAD). *Liber de rerum fossilium, lapidum et gemmarum, maxime figuris.* Tig. (Turin), 1565.

GIMMA (D. GIACENTO). *Della storia naturale delle gemme, delle pietre e di tutti minerali, ovvero della fisica sotteranea.* Napoli (Naples), 1730.

GINANNI (FANTUZZI M.). *Osservazioni geognostiche sul coloramento di alcune pietre e sulla formazione di un agate nel museo Ginanni di Rivenna.* 1857.

GIPPS (G. G. DE). *Occurrence of Australian opal.* [In a paper read before the Australian Institute of Mining Engineers, 1898.]

GLOCKER (ERNST FRIEDRICH). *De gemmis Plinii, imprimis de topazio.* Vratislaviæ (Breslau), 1824.

GOEPERT (H. R.). *Ueber pflanzenähnliche Einschlüsse in den Chalcedonen.* 1848.

Gold and Gems. Mawe's Travels in the Brazils. 1812.

GORCEIX (H.). *Les diamants et les pierres précieuses du Brésil.* [In Comptes Rendus, 1881, p. 981; also in Rev. Sci., xxix., 1882, p. 553.]

——. *Études des minéraux qui accompagnent le diamant dans le gisement de Salabro (Brésil).* [In Bull. Soc. Min. Francais, vii., 1884, p. 209.]

GRATACAP (LOUIS P., A.M.). *Curator, Mineralogy, American Museum of Natural History. A Vade Mecum Guide to Mineral Collections.* New York.

——. *The Collection of Minerals.* Supplement to American Museum of Natural History Journal, vol. ii., No. 2. February, 1902. Guide Leaflet No. 4 to the Museum.

GREGOR (WILLIAM). *An analysis of a variety of the corundum.* [In Nicholson's Journal, iv., 1803, p. 209.]

GREVILLE (CHARLES). *On the corundum stone from Asia.* [In Phil. Trans Abr., xviii., 1798, p. 356; and Nicholson's Journal, ii., p. 477.]

GRIFFITHS (A. B.). *On the origin and formation of the diamond in nature.* [In Chemical News, xlvi., 1882, p. 105.]

GROTH (P.). *Grundriss der Edelsteinkunde.* Leipzig, 1887.

GRONOVIUS (J.). *Gemmæ et Sculpturæ antiquæ de pictæ ab Leonardo Augustino Senensi.* 2 vols. in one. Franequeræ (Franecker), 1694.

GRÜNLING (FR.). *Über die Mineral vorkommen von Ceylon.* [In Zeits. Krystallographie, xxxiii., 1900, p. 209.]

GÜTHE (J. M.). *Ueber den Asterios-Edelstein des Caius Plinius Secundus; eine antiquarisch-lithognostische Abhandlung.* München (Munich), 1810.

GUYTON-MORVEAU (B. L.). *On the singular crystallization of the diamond.* [In Nicholson's Journal, xxv., 1810, p. 67.]

——. *Account of certain experiments and inferences respecting the combustion of the diamond and the nature of its composition.* [In Nicholson's Journal, iii., p. 298.]

HABDARRAHAMUS (ASIUTENSIS ÆGYPTIUS). *De proprietatibus ac virtutibus medicis animalum, plantarum ac gemmarum.* [Ex Arab. Lat. redd. ab Abrahamo Ecchellensi.] Paris, 1647.

HABERLE (C. C.). *Beobachtungen über Gestalt der Grün- und Keimkrystalle des schorlartigen Berylls, un dessen übrige oryctognostische und geognostische Verhältnisse.* Erfurt, 1804.

HAECKEL (E.). *A visit to Ceylon.* London, 1883.

HAIDINGER (W.). *Ueber den Pleochroismus des Amethysten.* Wien (Vienna), 1846.

——. *Ueber eine neue Varietät von Amethyst.* [In Denkschr. Akad. Wien, 1849.]

——. *Pleochroismus und Krystallstructur des Amethystes.* [In Ber. Akad. Wien, 1854.]

——. *Der für Diamant oder noch werthvolleres ausge-*

gebene Topas des Hern Dupoisat. [In Ber. Akad. Wien, 1858.]

HAMLIN (A. C.). *The tourmaline.* Boston, 1873.

——. *Leisure hours among the gems.* 1884.

HANNAY (J. B.). *On the artificial formation of the diamond.* [In Chemical News, 1880, p. 106.]

——. *Artificial diamonds.* [In Nature, xxii., 22, 1880, p. 255.]

HASSE (J. H. F.). *Der Aufgefundene Eridanus, oder neue Aufschlüsse über den Ursprung des Bernsteins.* Riga, 1769.

HAÜY (RENÉ JUST). *Traite de la minéralogie.* Paris, 1780.

——. *Mémoire sur les topazes du Brezil.* [In Ann. Mus. d'Hist. Nat., Paris, 1802.]

——. *Observations sur les Tourmalines, particulièrement sur celles qui se trouvent dans les États Unis.* [In Mémoire du Muséum, Paris, 1815.]

——. *Traite des caracteres physiques des Pierres précieuses, pour servir a leur determination lorsqu'elles sont tailles.* Paris, 1817.

HELMKACKER (R.). *On the Russian diamond occurrences.* [In Eng. and Min. Jour., Oct. 28, 1898.]

HOBBS (W. H.). *The diamond field of the Great Lakes.* [In Jour. of Geol., vii., 1899, no. 4.]

HERMES TRISMEGISTUS. *Tabula smaragdina vindicata.* 1657. An alchemical treatise.

HERTZ (B.). *Catalogue of Mr. Hope's collection of pearls and precious stones, systematically arranged and described.* London, 1839.

HESSLING (TH. VON). *Die Perlmuschel und ihre Perlen.* Leipzig, 1859.

HILLER (M.). *Tractus de Gemmis xii., in Pectorali Pontificis Hebræorum.* Tubingen, 1698.

HINDMARSH (R.). *Precious stones, being an account of the stones mentioned in the Sacred Scriptures.* London, 1851.

Histoire des Joyaux et des principales Richesses de l'orient et de l'occident. Geneve (Geneva), 1665.

History of Jewels. London, 1671.

HODGSON (JOHN). *Dissertation on an ancient cornelian.* [In Archæol., ii., 1773, p. 42.]

HOLCOMB (WILLIAM HARTLEY). *Precious gems and commercial minerals. . . .* San Diego: Press of Frye, Garrett & Smith, [190–?] 28, [4], pp. Illustrations. 12°.

HOLLANDUS (I.). *Opera mineralia et vegetabilia.* Arnhem (Arnheim), 1617.

HUDLESTON (W. H.). *On a recent hypothesis with respect to the diamond rock of South Africa.* [In Min. Mag., 1883, p. 199.]

Identification of Gems. [In Mineral Industries (annual), 1898, p. 278.]

JACOB (P. L.). *Curiosités des sciences occultes; alchimie, médecine chimique et astrologique, talismans, amulettes, baguette, divinatoire, astrologie, chiromancie, magie, sorcellerie, etc.* Paris, 1885.

JACOBS (H. and N. CHATRIAN). *Monographie du diamant.* Paris, 1880. A second edition in 1884.

JANETTAZ (N. and E. FONTENAY, EM. VANDERHEGEN, and A. COUTANCE). *Diamant et pierres précieuses.* Paris, 1880.

JANNETAZ (N.). *Les diamants de la couronne.* [In Science et Nature, 1884.]

JENNINGS (H.). *The Rosicrucians.* London, 1870. Another edition, 2 vols., in 1887. Contains some references to the mystical lore of gems.

JEFFRIES (DAVID). *Treatise on diamonds and pearls, in which their importance is considered, plain rules are exhibited for ascertaining the value of both, and the true method of manufacturing diamonds is laid down.* London, 1750.

——. *Traite des diamants et des perles.* Paris, 1753.

——. *An abstract of the treatise on diamonds and pearls, by which the usefulness to all who are in any way interested in these jewels will sufficiently appear, and therefore addressed to the nobility and gentry of this kingdom, and to the traders in jewels.* London, 1754.

JOHN (J. F.). *Naturgeschichte des Succins, oder des sogenannten Bernsteins.* Köln (Cologne), 1816.

JONES (W.). *Treasures of the earth, or mines, minerals, and metals.* London, 1879.

——. *Precious stones, their history and mystery.* London, 1880.

——. *Finger-ring lore.* London, 1890.

JONSTONUS (JOHANNES). *Notitia Regni Vegetabilis et Mineralis.* Lipseæ (Leipzig), 1661.

——. *Thaumatographia Naturalis.* Amsterdam, 1632.

JOSEPHUS. *Antiquatum Judaicarum.* [Translated from the Greek by W. Whiston.] London, 1737. In book iii., chap. viii., is an account of the marvellous properties of the stones in the breast-plate of the high priest.

JUDD (J. W. and W. E. HIDDEN). *On the occurrence of ruby in North Carolina.* [In Min. Mag., 1889, p. 139.]

JUTIER (——). *Exploitation du diamant dans la colonie du cap.* [In Compt. Rendus Soc. Industr. Min. St. Etienne, p. 84.]

Juwelier, Der Aufrichtige, oder Anweisung aller Arten Edelsteine, Diamenten, und Perlen zu erkennen, nebst einer aus dem Englischen uebersetzten Abhandlung von den Diamanten und Perlen. Frankfurt, 1772.

KAHLER (M.). *De Crystallorum Generatione.* Upsal (Upsala), 1747.

KALM (P.). *Några Kanne marken til nyttiga mineraliens eller ford och Baigarters upfinnande.* Aboæ (Abo), 1756.

Key to precious stones and metals. London, 1869.

KING (C. W.). *Antique gems.* London, 1860.

——. *The natural history of precious stones and of the precious metals.* London, 1867.

——. *The natural history of gems or decorative stones.* London, 1867.

——. *Handbook of engraved gems.* London, 1885.

KING (G. F.). *Topaz and associated minerals at Stoneham, Maine.* [In Am. Jour. Science, xxvii., 1884, p. 212.]

Kirani Kiranides et ad eas Rhyakini koronides, sive mysteria Physico-Medica. London, 1685.

[illegible] [illegible]. [illegible] in III [illegible] [illegible]). [illegible] [illegible] [illegible] in French.

[illegible] [illegible]. [illegible] for the [illegible] of gems. New York, [illegible].

[illegible] [illegible]. [illegible] of the [illegible]. [In [illegible] Journal [illegible].]

[illegible] [illegible]. [illegible] Geschichte [illegible] Berlin, [illegible].

[illegible] ——. [illegible]. Berlin, 1873.

——. *Die Edelsteine.*

[illegible] (K. E.). *[illegible] der Edelsteinkunde.* Leipzig, [illegible].

[illegible] (F. L. A. von). *[illegible] zur Gemmenkunde.*

——. *[illegible] über [illegible] und [illegible].* Braunschweig (Brunswick), [illegible].

KOKSCHAROW (N. von). *Materialien zur Mineralogie Russlands.* St. Petersburg. Eleven vols. and atlas. Begun in [illegible] and the parts issued from time to time. Contains mineralogical descriptions of gem minerals of the Russian Empire.

KÖNIG (EMANUEL). *Regnum minerale, physice, medice, anatomice, alchymice, analogice, theoretice et practice investigatum.* Basil (Basle), 1639.

KÖNNERITZ (L. VON). *Mittheilung mannichfaltiger Versuch Edelsteine Kunstgemeas zu schleifen.* Weimar, 1841.

KRAUSE (T. H.). *Pyrgoteles, oder die edeln Steine der alten in Bereiche der Natur, etc.* Halle. 1856.

KUNZ (G. F.). *Precious stones.* [In Mineral Resources of the United States. Issued annually by the United State Geological Survey.]

——. *Precious stones.* [In Appleton's Physical Geography.]

——. *The gems in the National Museum.* [In Popular Science Monthly, April, 1886.]

——. *Precious stones, gems, and decorative stones in Canada and British America.* [Ann. Rept. Geol. Survey of Canada. Ottawa, 1888.]

KUNZ (*Cont.*). *The fresh-water pearls and fisheries of the United States.* [In Bulletin of the U. S. Fish Commission, 1897, p. 375.]

——. *Gems and precious stones.* New York, 1890.

——. *Folk-lore of precious stones.* 1894. *A catalogue of specimens exhibited in the Department of Anthropology, World's Columbian Exposition, Chicago, 1893.*

——. *Sapphires from Montana, with special reference to those from Yogo Gulch, in Fergus County.* [In Am. Jour. Science, iv., 1897, p. 417.]

——. *Precious stones, minerals, etc.: original studies upon.* New York, 1902. [263] pp. Illustrations. Plate. Map. 8°.

——. *The production of precious stones in the United States.* Washington, 1900. [218] pp. Colored plates. Tables. 4°.

——. *Catalogue de la collection de pierres précieuses d'origine étrangère, exposée par la maison Tiffany & Co. pour le Museum d'histoire naturelle de New York. . . . Exposition universalle de 1900.* New York: Tiffany et Co. [1900]. 46 pp. 4°.

——. *Collection of pearls and the shells in which they are found in the brooks, rivers, and on the coasts of the United States.* U. S. section. Exposition universalle, Paris, 1900. New York, Paris [etc.]: Tiffany & Co. [1900]. [16] pp. 12°. (With his Catalogue de la collection de pierres précieuses . . . d'origine étrangére. New York, [1900]. 40°.

——. *Gems, jewellers' materials, and ornamental stones of California.* Sacramento: W. W. Shannon, Superintendent State Printing, 1905. 171 pp. Illustrations. Map. 8°. (California. State Mining Bureau. Bulletin No. 37.)

——. *Natal stones; sentiment and superstition connected with precious stones.* [9th ed.] New York: Tiffany & Co., [c 1902]. 30 pp. 16°.

KUNZ and STEVENSON (DR. GEORGE FREDERIC KUNZ and DR. CHARLES HUGH STEVENSON). *The Book of the Pearl.* Illustrated in colour tint, black and white, and photogravure. Royal octavo, $12.50. The Century Co., New York City. 1908.

LABARTE (M. JULES). *Handbook of the arts of the Middle Ages and Renaissance, as applied to the decoration of jewels, etc.* London, 1855.

LACAZE (DUTHIERS H.). *Histoire Naturelle du Corail, Organisation, Reproduction, Pêche en Algérie, Industrie, etc.* Paris, 1864.

LAET (JOHN DE). *De Gemmis et Lapidibus Libri II., Quibus præmittitur Theophrasti Liber; de Lapidibus Gr.Lat., cum annotationibus.* Ludg. Bat. (Leyden), 1647.

LANCON (H.). *L'Art du Lapidaire.* Paris, 1830.

LANGIUS (JOHANNES). *Epistolæ medicinales.* Lugd. (Leyden), 1557.

Lapidum Pretiosorum usus magicus, sive de sigillis. [In Harleian Manuscripts.]

LAUNAY (L. DE). *Les diamants du Cap.* Paris, 1897.

LEA (ISAAC). *Inclusions in gems.* [In Proc. Acad. Nat. Science, Philadelphia.]

——. *Further notes on "inclusions" in gems, etc.* Philadelphia: Collins, printer, 1876. 11, [1] pp. Plates. 8°. "Extracted from the Proceedings of the Academy of Natural Sciences of Philadelphia."

LEISNERUS (GOTT. CHRIST.). *De Coralliorum Natura, Præparatis et Usibus.* Wittembergæ (Wittemberg), 1720.

LEMNIUS (LEVINUS). *Occulta Naturæ Miracula.* Antwerp, 1567.

LENK (J.). *Neue Entdeckung eines Steines Serpentin-Agat.* Wien (Vienna), 1802.

LEONARDUS (CAMILLUS). *Speculum Lapidum.* Venet. (Venice), 1502.

——. *Tratto delle Gemme che produce la Natura; traduzione di M. Ludovico Dobe.* 1565.

——. *The mirror of stones, in which the nature, generative properties, virtues, and various species of more than 200 different jewels, precious and rare stones are distinctly described.* London, 1750.

LEWIS (H. C.). *Genesis and matrix of the diamond.* London, New York, and Bombay, 1897.

LIBAVIAS (A.). *Alchemia.* Frankfurt, 1597.

Liber Hermetis, tractans de 15 Stellis et de 15 Lapidibus et de 15 Herbis et de 15 Figuris. [In Harleian Manuscripts.]

LIVERSEDGE (A.). *On the occurrence of diamonds in New South Wales.* [In Minerals of New South Wales, London, 1888.]

LOEWM (——). *Ueber den Bernstein und die Bernstein-Fauna.* Berlin, 1850.

LONINSER (G.). *Die Marmaroscher Diamanten.* Presberg, 1856.

LÖSCH (A). *Ueber Kalkeisengranat (Demantoid) von Syssertzk am ural.* [In Neues Jahrbuch, 1879, p. 785.] Description of locality, occurrence, etc., of the green garnet (demantoid) used in jewelry.

LOUIS (H.). *The ruby and sapphire deposits of Moung Klung, Siam.* [In Min. Mag., 1894, p. 276.]

LUCRETIUS (——). *De Rerum Natura.*

LULLIUS (RAYMUNDIS). *Lebelli aliquot chemici, etc.* Basileæ (Basle), 1600. [See p. 819: "De compositione gemmarum et lapidum preciosorum."]

MAKOWSKY (A.). *Ueber die Diamanten des Kaplandes auf der Weltaustellung in Wien.* [In Verh. Nat. Ver Brünn, xii., p. 16.]

MALLET (F. R.). *On sapphires recently discovered in the Northwest Himalayas.* [In Rec. Geol. Surv. India, xv., 1881, p. 138.]

MANDEVILLE (JOHN). *Le Grande Lapidaire, où sont déclarez les noms de Pierres orientales, avec les Vertus et Propriétés d'icelles, et iles et pays où elles croissant.* Paris, 1561.

MARBODAEUS (GALLUS). *De Gemmarum Lapidumque pretiosorum formis atque viribus opus culum.* Colon (Cologne), 1593.

——. *De Lapidibus pretiosis Enchiridion, cum Scholiis Pictorii.* Wolfenbültelæ (Wolfenbültel), 1740.

MARIETTE (P. J.). *Traité des Pierres gravées.* Paris, 1750.

MARSHALL (W. P.). *Notes on the Great Kimberley Diamond Mine.* [In Midl. Nat., vii., p. 93.] Marlborough gems. Gemmarum Antiquarum Delectus ex præstantioribus desumptus, quæ in Dactyliothecis Ducis Marlburiensis conservantur, 1845.

MARTIN (K.). *Notizen über Diamanten.* [In Zeits. deutsch. geol. Gesells., xxx., p. 521; plate.] A crystallographic study of the diamonds in the Leyden Museum.

MASKELYNE (N. S.). *Artificial diamonds.* [In Jour. Soc. Arts, xxvii., p. 289.]

MASON (F.). *Burma: Its people and productions.* London, 1882. In 2 vols., i., geology and mineralogy.

MAWE (JOHN). *A treatise on diamonds and precious stones, including their history, natural and commercial. To which is added some account of the best method of cutting and polishing them.* London, 1813.

——. *Travels in the interior of Brazil, particularly in the gold and diamond districts of that country.* London, 1812.

MEINEKE (J. L. G.). *Ueber den Chrysopras und die denselben begleitenden Fossilien in Schlesien.* Erlangen, 1805.

Metropolitan Museum of Art. The Collection of Engraved Gems. Introduction to and description of The Johnston Collection at the Museum. Metropolitan Museum of Art. The Heber R. Bishop Collection of Jade and other Hard Stones. Descriptive matter of pertinent Mineralogy, Archæology, and Art.

MEUNIER (S.). *Composition et origine du sable diamantifére de Du Toits Pan (Afrique-Australe).* [In Comptes Rendus, lxxxiv., p. 250.]

MIDDLETON (J. HENRY). *The engraved gems of classical times, with a catalogue of the gems in the Fitz-William Museum.* Cambridge: University press, 1891. xvi., 157, xxxvi., pp. Illustrations. Plates 4°.

——. *The Lewis collection of gems and rings in the possession of Corpus Christi college, Cambridge; with an introductory essay on ancient gems.* London: C. J. Clay and Sons, 1892. 93 pp. Illustrations. 8°.

MILES (C. E.). *Diamonds.* [In Trans. Liverpool Geol. Soc., ii., p. 92, 1882.]

M. L. M. D. S. D. *Dénombrement, Faculté et Origine des Pierres précieuses.* Paris, 1667.

MÖBIUS (K.). *Die echten Perlen.* Hamburg, 1857.

MORALES (G. DE). *Libro de las Virtudes y Propriedades maravillosas de las Piedras preciosas.* Madrid, 1605.

MORGAN (SYLVANUS). *The Sphere of Gentry.* 1661. Contains an account of the heraldic meaning of gems.

MORRIS (J.). *Gems and precious stones of Great Britain.* 1868.

MORTIMER (CROMWELL). *Remarks on the precious stone called the turquois.* [In Phil. Trans. Abr., viii., p. 324.]

MÜLLER (J.). *Nachricht von den in Tyrol entdeckten Turmalinen, oder Aschenziehern, von Ignaz Edeln von Born.* Wien (Vienna), 1787.

MURRAY (J.). *Memoir on the diamond.* London, 1839.

MURRAY (R. W.). *Diamond fields of South Africa.* [In Jour. Soc. Arts, xxix., p. 370.]

NATTER (L.). *A treatise on the ancient method of engraving precious stones compared with the modern.* London, 1754.

Natural Magick, in twenty books, wherein are set forth all the riches and delights of the natural sciences, with engravings. London, 1658. An English trans. of Porta's Magiæ Naturalis.

NANMANN (KARL FRIEDRICH). *Lehrbuch der reinen und angewandten Krystallographie.* Liepzig: F. A. Brockhaus, 1830. 2 vols. 905 diagrams on xxxix. folded plates. 8°.

NICOLS (THOMAS). *A lapidary, or history of pretious stones; with cautions for the undeceiving of all those that deal with pretitious stones.* London, 1754.

——. *Arcula Gemmea; or the Nature, Virtue, and Valour of Precious Stones, with cautions for those who deal in them.* Cambridge, 1652.

——. *Gemmarius Fidelis, or the Faithful Lapidary; experimentally describing the richest Treasures of Nature, in an Historical Narrative of the several*

Natures, Virtues, and Qualities of all Precious Stones, with a Discovery of all such as are adulterate and counterfeit. London, 1659.

NORTHRUP (H. D.). *Beautiful gems.* 1890.

OCHTCHEPKOFF (J. W.). *Qui a découvert le Diamant dans les Montes Ourals?* [In Bul. Soc. Oural. Sci. Nat., vii., p. 87, 1884.]

Opals (Australian). [In Iron, xxii., p. 490, 1883.]

ORPEN (G.). *Stories about famous precious stones.* 1890.

ORPHEUS. *Hymni et de Lapidibus. Gr. Lat., curante A. C. Eschenbachio; accedunt H. Stephani notæ.* Traj. ad Rh. (Cologne), 1689.

ORTON (J.). *Underground treasures.* Philadelphia, 1881.

PAGE (D.). *Economic Geology.* London, 1874.

PARACELSUS (PHILIPPUS AURELIUS THEOPHRASTUS). *Nine books on the nature of things;* into English by J. F. London, 1650.

——. *Of the chymical transmutation, genealogy, and generation of metals and minerals* [tr. by R. Turner]. London, 1657.

PARROT (——). *Notices sur les Diamants de l'Oural.* [In Mem. de l'Acad. Imp., St. Petersburg, 1832.]

PARTSCH (P.). *Beschreibendes Verzeichniss einer Sammlung von Diamanten und der zur Bearbeitung derselben nothwendigen apparate.* Wien (Vienna), 1822.

PAXMAN (J. N.). *The diamond fields of South Africa.* [In Eng. Min. Jour., xxxv., p. 382.]

——. *On the diamond fields and mines of Kimberley, South Africa.* [In Proc. Inst. Civil Eng., lxxiv., p. 59.]

PAXTON (J. R.). *Jewelry and the precious stones.* [By Hipponax Roset, pseudon.] Philadelphia, 1856.

PETZHOLDT (M.). *Beiträge zur Naturgeschichte des Diamants.* Dresden und Leipzig, 1842.

PHILOSTRATUS. *De Vita Apolonii.*

PIERERUS (G. P.). *Lazulus, Dissertatio chymico, medica.* Argentorati (Strasburg), 1668.

PINDER (——). *De Adamante Commentatio Antiquaria.* Berlin, 1829.

PIRSSON (L. V.). *On the corundum-bearing rock from Yogo Gulch, Montana.* [In Am. Jour. Science, iv., 1897, p. 421.]

PLINY. *Historia Naturalis C. Plinii secundi.* First issued A.D. 77. The work is divided into 37 books, and these into short chapters; the last 5 books treat particularly of gems and other minerals.

PLUCHE (ANTOINE NOËL DE). *Spectacle de la Nature.* Paris, 1732–39.

PLUMMER (J.). *Australian localities of diamond.* [In Watchmaker, Jeweller, and Silversmith, xxiv., 1898.]

PLYTOFF (G.). *Divination, calcul, des probabilités oracles et sorts, songes, graphologie, chiromancie, phrénologie, physiognomie, cryptographie, magie, kabale, alchimie, astrologie, etc.* Paris, 1891.

POLE (W.). *Diamonds.* [In London Archæol. Trans., 1861.]

PORTA (JOHN BAPTIST). *Magiæ Naturalis.* Porta [born 1538, died 1615] published the first edition of this work in 1553, when he was but 15 years old. It contains much concerning the mystical properties of gems. The work also contains a description of the camera obscura.

——. *A method of knowing the inward virtues of things by inspection.* 1601.

——. *De Distillationibus.* Romæ (Rome), 1608.

PORTALEONE (ABRAHAM). *Shilte Haggeborim* [*The Shields of the Mighty*]. Mantua, A.M. 5372 (A.D. 1612).

POTT (M. J.). *Lithogeopnosie ou examen chymique des Pierres et des Terres en général et de la Topaze et de la stéatite en particulier.* Paris, 1753.

POUGENIEFF. *Precious Stones.* Russian, with 2 coloured plates and numerous woodcuts. St. Petersburg, 1888.

POUGET (N.). *Traité des Pierres précieuses, et de la manière de les employer en parure.* Paris, 1762.

PRATT (J. H.). *Notes on North Carolina minerals.* [In Journal Elisha Mitchell Scientific Society, xiv., part 2, p. 61, 1898]. Describes occurrence of emerald.

PRATT (J. H., and W. E. HIDDEN). *Rhodolite, a new va-*

riety of garnet. [In Am. Jour. Science, v., 1898, p. 293; also vi., 1898, p. 463.]

Precious stones of the Bible; descriptive and symbolical. 1878.

Precious stones, cutting and polishing of. [In Mineral Industries (annual), p. 229, 1899.]

PRINZ (W.). *Les enclaves du Saphir, du Rubis, et du Spinelle.* [In Bul. Soc. Belge. Microsc., 1882.]

PSELLUS (MICHAEL CONSTANTINUS). *Le Lapidum Virtutibus.* Lugundi Batavorum (Leyden), 1795.

RAGOUMOVSKY (——). *Distribution Technique des Pierres précieuses, avec leurs caractères distinctifs.* Vienne (Vienna), 1825.

RAMBOSSON (——). *Les Pierres précieuses.*

RANTZOVIUS (HENRY). *De Gemmis scriptum olim a poeta quodam non infeliciter carmine redditum et nunc primum in lucem editum.* Leipzig, 1585. A manuscript on the properties and effects of precious stones attributed to "Evax, a King of the Arabs."

RAVIUS (S. F.). *Specimen arabicum, continens descriptionem et excerpta libri Achmedis Teifaschii "De Gemmis et Lapidibus Pretiosis."* Trajetum ad Rhenum (Leyden), 1784.

REYNAUD (J.). *Histoire élémentaire des minéraux usuels.* Paris, 1867.

ROBERTSON (J. K. M.). *The occurrence of opals in central Australia and Queensland.* [In Chem. News, lxv., 1882, pp. 95. 101.]

——. *On the occurrence of opals in the colony of Queensland.* [In Proc. Phil. Soc. Glasgow, xiii., p. 427.]

RONTON (EDWARD). *Intaglio engravings, past and present.* London: G. Bell and Sons, 1896. xii., 117, (2) pp. Illustrations. 16°.

ROSENMÜLLER (E. F. C.). *Mineralogy of the Bible.* [Translated by Repp and Morren.] Edinburgh, 1840.

ROSS (W. A.). *Pyrology.* London, 1875.

——. *On the cause of the blue colour of sapphire, lazulite, and lapis lazuli; the green colour of emerald and the purple of amethyst.* [In Chem. News, xlvi., 1882, p. 33.]

ROTHSCHILD (M. D.). *Handbook of precious stones.* New York, 1890.

ROY (C. W. VAN). *Ansichten über Enstehung und vorkommen des Bernsteins, so wie praktische mittheilungen über den werth und die Behandlung desselben als Handelswäre.* Dantzig, 1840.

RUDLER (F. W.). *Agate and agate working.* [In Pop. Science Rev., i. (new series), p. 23.]

——. *Artificial diamonds.* [In Pop. Science Rev., iv., 1880, p. 136.]

——. *Diamonds.* [In Science for All, ii.]

——. *On jade and kindred stones.* [In Pop. Science Rev., iii., p. 337.]

RUDOLPH (A.). *Die edeln metalle und Schmucksteine, mit 87 Tabellen.* Breslau, 1858.

RUE (F. DE LA). *De Gemmis.* Parisii (Paris), 1547. Other editions: Ludg., 1622; Franc., 1626; Gron., 1626.

RUENS (F.). *De Gemmis aliquot, iis praesertim quarum Divus Joannes Apostolus in sua Apocalypsi notavit.* Paris, 1547.

RULANDUS (M.). *Medicina Practica.* Arg. (Strasburg), 1564.

——. *Lexicon Alchemiæ.* Frankfurt, 1661. First ed. dated 1612. The author, a physician to Rudolph II. of Germany, gives several receipts for the development of the occult and medicinal properties of gems.

RUSKIN (J.). *On the heraldic meaning of precious stones.* [In his lecture before the London Institute, Feb., 1876.]

SANDIUS (CHRISTOPHER). *On the origin of pearls.* [In Phil. Trans.: Abr., ii., p. 126, 1674.]

Sapphire. [In Mineral Industries (annual), p. 235, 1896.]

Sapphire Mines of Burma. [In Mineral Industries (annual), p. 239, 1896.]

Sapphire Mines in Siam. [In Jour. Soc. Arts, xxviii., p. 770.]

SARMENTO (J. C. DE). *An account of diamonds found in Brazil.* [In Phil. Trans.: Abr., viii., 1731, p. 503.]

SHAW (FREDERICK MOULTON). *Gems of the first water*

and how obtained. [2d ed.] Los Angeles, Cal.: F. M. Shaw and G. Bentley [1887]. 3–99 pp. 24°.

SCHINDLER (A. H.). *The turquoise mines of Nishapur, Khorassan.* [In Rec. Geol. Survey, India, xvii., p. 132.]

SCHMIDT (C. J.). *Das Wichtigste über dem Opal in Allgemeinen und über sein Vorkommen in Mähren im Besonderen.* [In Mitth. d. k. k. Mähr. Schles. Gesells., Brunn, 1855.]

SCHRAUF (A.). *Handbuch der Edelsteinkunde.* Wien (Vienna), 1869.

SCHULZE (H.). *Practisches Handbuch der Juwelierkunst und Edelsteinkunde.* Quedlinburg und Leipzig, 1830.

SCOT (REGINALD). *Discovery of witchcraft.* London, 1651. Contains several curious charms in which gems are used.

SCUDALUPIS (P. ARLENSIS DE). *Sympathia Septem ac Septem Selectorum Lapidum ad Planetas.* An alchemical or astrological work; among other curiosities it contains a list of stones "in sympathy with the seven planets."

SERAPION (J.). *De medicamentis tam simplicibus quam compositis.* Mediolanum (Milan), 1473.

SHELLY (F.). *Legends of gems.* New York, 1893.

SHEPARD (C. U., SR.). *Notice of corundum gems in the Himalaya regions of India.* [In Am. Jour. Science, xxvi., 1883, p. 339.]

SHEPSTONE (T.). *The geographical and physical characters of the diamond fields of South Africa.* [In Jour. Soc. Arts, xxii., 1874.]

SHIPTON (——). *Precious gems.* London, 1867.

SILLIMAN (B.). *Turquoise of New Mexico.* [In Proc. Am. Assoc., xix., p. 431; also Am. Jour. Science, xxii., 1880, p. 67.]

SLEVOGTII (J. H.). *De Lapide Bezoar.* Jenæ (Jena), 1698.

SMITH (M. A. H. CLIFFORD). *Jewellery.* G. P. Putnam's Sons, New York, 1908.

SOMMERVILLE (MAXWELL). *Engraved Gems.* Philadelphia, London, [etc.]: D. Biddle, 1901, 6, (2), 7–183 pp. Illustrations. Plates. 4°.

SOTTO (JS.). *Le Lapidaire du Quartorzième Siècle.* Wien (Vienna), 1862.

SPENCER (GEORGE). *Gemmarum antiquarum delectus; ex praestantioribus desumptus, quae in dactyliothecis ducis Marlburiensis conservantur.* Londini: Apud Joannen Murray, 1845. 2 vols. Plates. F°.

SPENER (J. J.). *De gemmis errores vulgares.* Lipsiæ (Leipzig), 1688.

SMYTH (H. W.). *Five years in Siam (1891–1896).* 2 vols. London, 1898.

SPEZIA (G.). *Sul colore del Zircone.* [In Atti R. Ac. Torino, xii., p. 37.] His experiments show that the colour is dependent upon the degree of oxidation of the contained iron.

STEINBECK (——). *Ueber die Bernstein-Gewinnung.* Brandenburg, 1841.

STERRETT (DOUGLAS B.). *The Production of Precious Stones in 1906.* Advance Chapter from Mineral Resources of the United States, Calendar Year, 1906. Department of the Interior—United States Geological Survey. Government Printing Office, Washington, D. C., 1907.

STREETER (E. W.). *Precious stones and gems.* London, 1877. [1882.]

——. *Great diamonds of the world.* London, 1892. [1898.]

SUTTON (A. L.). *Lingua gemmæ: cycle of gems.* 1894.

TAGORE (S. M.). *Mani-málá, or a treatise on gems.* 2 vols. Calcutta, 1879. Contains a bibliography of Sanskirt, Persian, Arabic, and other Oriental works on gems.

TASSIE (JAMES). *Descriptive catalogue of a collection of ancient and modern engraved gems, cameos, and intaglios of the most celebrated cabinets in Europe; cast in colored pastes, white enamel, and sulphur; arranged and described by R. E. Raspe, 1791.*

TASSIN (WIRT). *Descriptive catalogue of the collections of gems in the United States National Museum.* Washington: Government printing office, 1902. (2), 473–670 pp. Illustrations. Plates. 8°. "Reprinted from the Report of the United States National Museum for 1900." Bibliography, pp. 649–670.

The Mussel Fishery and Pearl Button Industry of the Mississippi River. United States Fish Commission Bulletin for 1898. Pages 289–314. Plates 65 to 85.

TAVERNIER (J. B.). *Voyages in Turquie, en Perse et aux Indes.* Paris, 1676.

——. *Account of diamond mines.* [In Pinkerton's Collection of Voyages, viii., 1811.]

TAYLOR (L.). *Precious stones and gems, with their reputed virtues.* London, 1895.

TAYLOR (N.). *On the Cudgegong diamond field, New South Wales.* [In Geol. Mag., iv., p. 399.]

TEIFASCITE (AHMED). *Fior di Pensieri sulle Pietre Preziose, opera stampata nel suo originale Arabo di Ant. Raineri.* Firenze (Florence), 1818.

TENNANT (J.). *Gems and precious stones.* [In Soc. of Arts, Lect., 1851–52.]

Tesoro delle Gioie, Trattato curioso. Venetitia (Venice), 1670.

THEOPHRASTUS. *History of stones, with the Greek text and an English version, and notes, critical and philosophical, including the modern history of gems described by that author,* by Sir John Hill. London, 1746.

Thousand (A) notable things on various subjects. London, 1814.

TIMBERLAKE.—*Discourse of the travels of two English pilgrims.* 1611. Contains, among others, an account of a great jewel used in conjuring.

TOLL (ADRIANUS). *Gemmarum et Lapidum Historia.* Lugduni (Leyden), 1636.

——. *Le Parfaict Joaillier, où Histoire des Pierreries, où sont amplement descrites leur naissance, juste prix, etc.* Lyon, 1644.

Traité des Pierres de Théophraste. [Translated from the Greek.] Paris, 1754.

TURNER (H. W.). *The occurrence and origin of diamonds in California.* [In American Geologist, 1899, p. 182.]

VALENTINUS (BASILUS). *Of natural and supernatural*

things, etc. [Translated from the Dutch by D. C.] London, 1670. An alchemical treatise containing several accounts of the occult and medicinal properties of gems. The German edition was issued at Eisleben in 1603.

VANE (G.). *The pearl fisheries of Ceylon.* [In Journal of the Ceylon Branch Royal Asiatic Society, x., 1887. Colombo.]

VEGA (GARCILASO DE LA). *History of the Incas.* [Various editions.]

VELTHEIM (A. F. VON). *Etwas über Memnons Bildsäule, Nero's Smaragd, Toreutik, und die Kunst der Alten in Stein und Glas zu schneiden.* Helmstadt, 1793.

——. *Etwas über das Onyx-Gebirge des Clesias und den Handel der Alten nach Ost-Indien.* Helmstadt, 1797.

VENETTE (NICHOLAS). *Traité des Pierres.* Amsterdam, 1701.

VETTERMANN (A.). *Kurze Abhandlung über einige der vorzüglichsten Classen der bunten oder gefärbten Edelsteine.* Dresden, 1830.

VOGEL (H. W.). *Spectralanalytische Notizen.* [In Ber. Deutsch. chem. Gesell., x., p. 373, 1887.] Examination of garnet, ruby, etc.

WASHBURN (HOWARD E.). *American Pearls.* The Ann Arbor Press, Ann Arbor, Mich.

WERNHER (——). *Die Gewinnung und Aufbereitung der Diamanten in Süd-Afrika.* [In Wochenschr. Deutsch. Ing.-Arch.-Ver., p. 365.]

WESTROPP (H. M.). *Manual of precious stones.* 1874.

WECKERUS (or WECKER). *Antidotæ speciales de Lapidibus minus pretiosis alterantibus.*

WILLIAMS (C. G.). *Researches on emeralds and beryls.* [In Chem. News, xxv., p. 256.] A purely chemical paper.

WILLIAMS (GARDNER F., General Manager De Beers Mines.) *The Diamond Mines of South Africa.* B. F. Buck & Co., New York, 1905.

WILLIMOT (C. W.). *Canadian gems and precious stones.* [In Ottawa Naturalist, Nov., 1891.]

WORLIDGE (T.). *A select collection of drawings from*

curious antique gems; most of them in the possession of the nobility and gentry of this kingdom; etched after the manner of Rembrandt. London: Printed by Dryden Leach, for M. Worlidge, 1768. (12), 48 pp. Plates. 4°.

ZEPHAROVITCH (V. VON). *Der Diamant, ein populärer vortrag.* Gratz, 1862.

ZERREMER (C.). *Anleitung zum Diamanten. Waschen aus Seifengebirge, Uferund Flussbett-Sand.* Leipzig, 1851.

——. *De Adamanti Dissertatio.* Lipsiæ (Leipzig), 1862.

GLOSSARY

ACICULAR. Needle-like.

ADAMANTINE. Very hard—as hard as steel. From Adamas (Greek); Adamanta (Latin), the lustre of the diamond.

AGGREGATES. Clusters or groups.

ALLUVIAL. Washing away rocks, soil, or other mineral material from one place and depositing the *débris* in another.

AMORPHOUS. Without form, shapeless.

AMULET. From hamalet (Arabian), to carry. A charm, or talisman, worn on the person to ward off disease, accident, or other harm.

ARBORESCENT. Resembling a tree in appearance.

ASTERIATED. Radiated, with rays diverging from a centre, as in a star—as exhibited by an asteriated or star sapphire.

AVICULIDAE. Wing-shells, or Pearl Oysters.

AXIS. Axes or planes of crystals or other minerals—as demonstrated in crystallography.

BABY. Trough or cradle in which gravel was washed for diamonds by early South African diamond-seekers.

BAHIAS. Diamonds from the Bahia district, Brazil.

BASE. "Foundation price of a one-grain pearl from which to reckon prices of pearls of other weights. The price of pearls is quoted by the grain and reckoned by the square; example: a two-grain pearl at three dollars base would be twice three dollars, or six dollars per grain 'flat'; and two grains at six dollars would be twelve dollars, the cost of the pearl." (From *Precious Stones* by W. R. Cattelle.)

BIREFRINGENCE. Double refraction of light of crystal minerals.

BIZEL. Portion of brilliant-cut diamond above the girdle.

BLEBBY. Blisters or bubbles in a crystal mineral

BLUE GROUND. Diamond-bearing clay of lower levels of South African diamond mines.

BLUE WHITE. Highest grade of South African diamonds.

BORT or BOART. Imperfectly crystallised form of diamond unfit for gems and used for pointing rock drills, for bearings of fine machinery and other technical uses.

BOTRYOIDAL. A surface presenting a group of rounded projections.

BRECCIA. A not wholly formed rock of angular fragments naturally cemented by lime or some other adhesive mineral substance or "binder."

BRILLIANT. A style of diamond-cutting with fifty-six facets, exclusive of table and culet.

BRITTLE. A mineral, when it may be readily broken by a blow.

BRITTLE. A stone that breaks, or parts of it separate into powder, when the attempt is made to cut it.

BROWNS. Eighth in list of principal trade terms in grading diamonds.

BRUTING. Polishing diamonds by rubbing one against another.

BUBBLES. Small hollow specks in the body of a gem.

BUILT-UP RUBY. Reconstructed ruby.

BYON. Brownish-yellow clay in which occurs corundum—rubies, sapphires, etc.

—— Ground adjacent to mother rock in which rubies have weathered out.

BYSSUS. Fibres, flaxy or silky in appearance, by which a mussel attaches its shell to wood or stone.

BY-WATERS. Yellow tinted diamonds.

CAPES. Diamonds with a yellowish tinge.

CAPILLARY. Hair-like.

CARAT. (Karat.) A unit of weight applied to precious stones verying in different trade centres. See table of weights of the carat in various localities in the Appendix.) The word carat is supposed to be derived from "Kuara," the bean-like fruit of an African tree reputed to have been used as a standard of weight for precious stones. Karat is used to indicate degrees of quality in gold.

CARBON. A tetrad (having four sides), non-metallic mineral element occurring in two crystalline forms, diamond and graphite, and one amorphous form, coal.

CARBON DIOXIDE. Carbonic acid gas; a colourless gas 1524 times as heavy as air and twenty-two times as heavy as hydrogen.

CARBON SPOTS. Opaque black spots in the body of a diamond.

CARBONADO. Brownish, black variety of diamond; large pebbles or masses of diamonds, nearly pure carbon. Carbonado was formerly chiefly found in great quantity—now decreasing—in Bahia diamond district, Brazil; used to point rock drills and, reduced to powder, for polishing diamonds.

CARBUNCLE. Garnet—sometimes, ruby, spinel, or other red gem—cut convex or *en cabochon:* there is no such specific mineral.

CAT'S-EYE. A term applied to gem minerals which, when cut convex (*en cabochon*), display a band of light, usually across inclusions of parallel fibres of asbestos; name derived from resemblance to the eye of a cat.

CEYLON RUBY. A ruby having a pink tint.

CHALCEDONY PATCHES. Milk-like semi-transparent patches which sometimes occur as faults in rubies.

CHANGE OF COLOURS. Manifested in minerals like Labradorite, where the colours change as the stone is turned.

CHATOYANCY. Changeable or undulating lustre or colour, as displayed by a cat's-eye.

CHIPS. Cleavage of diamonds of smallest fractions of a carat in use.

CLATERSAL. Diamond splints, which are converted into diamond powder by crushing.

CLEAN. Free from interior flaws.

CLEAVAGE. Direction within a crystal along which there is minimum cohesion; diamond crystals which require cleaving; pieces cleaved from the crystal.

Cleaving. Splitting a crystal in a direction in which it may most easily be done—along the grain.

Close Goods. Pure stones, of desirable shapes; highest class of South African diamonds, as assorted at Kimberley.

Clouds. Muddy or cloudy patches of any colour in a stone which, when brought to the surface by cutting, are ineradicable. "Flat, subtransparent blotches along the grain of a stone."—Cattelle.

Colour-Play. (Play of Colours.) Prismatic colours produced by dispersion of light.

Colour Range. A statement of the various colours exhibited by different specimens of a mineral.

Combustibility. A quality possessed by the diamond only, among gems.

Concentrates. Gem or mineral ore or ground reduced by mechanical or chemical processes to its minimum in bulk or weight.

Conchoidal. Shell-like fracture of any mineral.

Concretions. Mechanical aggregation, or chemical union of particles of mineral forming balls or irregularly shaped nodules in strata of different material.

Conglomerate. Pebbles or gravel bound together naturally by a silicious, calcareous, or argillaceous cement.

Corundum. Crystallised alumina—rubies, sapphire, etc.

Cradle. Trough in which, by a rocking motion, placer miners wash auriferous or gem gravels.

Crystallography. The science which describes or delineates the form of crystals.

Crystals. Trade term for fourth grade cut diamonds; colourless diamonds.

Culasse. Portion of brilliant-cut diamond below the girdle.

Culet. (Or Collet). Bottom facet of brilliant parallel to the girdle.

Curator. One to whose official care is entrusted a department—as of mineralogy—in a museum.

Diamond. The mineral gem alone composed of pure carbon; crystallises in the isometric, or cubic, system; combustible, it can be totally consumed, disappearing in carbonic acid gas, when burned between the poles of a powerful electric battery.

DIAPHANEITY. The property of transmitting light.

DICHROISM. A property of all doubly refractive stones, of which the two images revealed by an instrument called dichroiscope appear in different colours.

DICHROSCOPE. An instrument designed to exhibit the two complementary colours of polarised light—the dichroism of crystals.

DISPERSION. The power which decomposes a ray of common white light in its passage through a transparent medium and splits it into the various colours of which it is composed.

DODECAHEDRON. A geometrical form in the isometric or cubic system applied to crystallography; a solid figure of twelve equal sides, each a regular pentagon —of five equal sides and angles.

DOLOMITIC. Pertaining to dolomite, a brittle, translucent mineral of various colours and a vitreous lustre.

ERUPTIVE. Minerals of volcanic origin in geological formations.

FACET. One of the small planes which form the sides of a natural crystal, or of a cut diamond or other gem.

FALES. Stones of two, or more, differently tinted strata.

FALSE COLOUR. Effect of "False Stones."

FANCY. A term that has been applied to semi-precious stones prized for other qualities than intrinsic value.

FAULT. Anything within, or on the surface of, a precious stone which detracts from its beauty or value; obvious examples are inclusions of foreign bodies and patches of a different colour or shade from the body of the gem.

FEATHERS. White subtransparent lines in the body of a stone.

FEMININE. Rubies of a pale tint.

FERROUS. Any mineral substance having a considerable portion of iron in its composition.

FIRE. Term applied to the lustre and brilliancy of gems, pre-eminently the diamond, and secondarily the opal.

First Bye. (First By-water.) Diamond exhibiting a faint greenish tinge.

First Water. Diamonds so pure and colourless that they can scarcely be distinguished from water when immersed in it.

Fish-Eye. A diamond cut too thin to present the maximum effect of brilliancy.

Flat Ends. Thin cleavages from the faces of a diamond crystal.

Flats. Thin, flat pieces of diamond crystal.

Flaw. A crack, defect, fault, fissure, or other structural imperfection in a gem.

Fluorescence. The phenomenal quality exhibited by some gems of showing one colour in transmitted light and another in reflected light; fluorite, from which the word is derived, is a striking example.

Flux. To melt, to fuse. As a noun, a fluid or substance which may be used to fuse some other material.

Fracture. Breaking a gem otherwise than the lines of cleavage.

Gem Colour. The most desirable colour for a stone.

Gemology. A word coined to supply a specific name for the science of gems.

Glassies. Octahedral diamond crystals (transparent).

Glassy. Applied to diamonds which lack brilliancy.

Golconda. Ancient and famous group of diamond mines on the Kistna River, India, where were found the Koh-i-noor and other world-famous diamonds.

Golcondas. Diamonds from India.

Grain Marks. Lines on the facet surfaces, the result of imperfect polishing.

Grainers. Diamonds which in weight will correspond to fourths of a carat; a diamond weighing one half a carat is a two-grainer; one weighing three quarters is a three-grainer; a diamond of one carat in weight is a four-grainer.

Granitic. Like, or of, granite.

Granular. Composed of or resembling granules or grains.

Harlequin. Most beautiful variety of opal.

Hemihedral. Having only half the planes or facets which a symmetric crystal of the type to which it belongs would possess; a crystal wanting some of its planes. (The hemihedral form in crystallography produces or aids the phenomena of pyroelectricity.)

Hexagonal. Of the form of a hexagon; having six sides or angles.

Hydrostatics. Pertaining to the principles of the equilibrium of fluids.

Inclusions. Foreign substances within the body of a transparent mineral.

Indian-Cut. A style of diamond-cutting usually of Indian or other Oriental origin in which the table is usually double the size of the culet; such stones are generally recut for European or American requirements.

Iridescence. Descriptive of prismatic colours appearing within a crystal.

Isometric. The cubic system in crystallography.

Jagers. Bluish-white diamonds of modern cut; originally diamonds from the Jagersfontein mine.

Jig. (Jigger; Pulsator.) A riddle or sieve shaken vertically in water to separate ore or gem gravel or ground into strata.

Knife-Edge. The girdle of a brilliant cut to a sharp edge and polished.

Knots. Conditions found in diamonds as in wood, and troublesome to the lapidary.

Lapidary. One who cuts, polishes, or engraves precious stones.

Light Yellow. Seventh grade diamonds.

Lumpy. Stones cut thick.

Lustre. The optical character of a gem, dependent upon that portion of the light falling upon it which is reflected from the surface. Degrees of lustre: splendent, shining, glistening, glimmering. Kinds of lustre: metallic, vitreous or glassy, adamantine (the diamond's lustre), silky, satiny, pearly, nacreous, greasy, waxy, resinous.

Maacles. Flat triangular diamond crystals or twin stones.

Macled. Twinned crystals.

Masculine. A term applied to rubies of an intensely red hue.

Matrix. The portion of rock in which a mineral is embedded. Gem minerals are sometimes cut together with a portion of the matrix and the matrix itself is sometimes cut and mounted like gems.

Melange. Diamonds of mixed sizes.

Melee. Small diamonds.

Metallurgy. The art of separating metals from their ores or from impurities; smelting, reducing, refining, amalgamating, alloying, parting, brazing, plating, etc.

Mineralogy. A science treating of those natural inorganic products of the earth which possess definite physical and chemical characters.

Monoclinic. Inclining in one direction.

Monoclinic System. Having two of the axial intersections rectangular and one oblique; having the lateral axes at right angles to one another, one of them being oblique to the vertical axis and the other at right angles to it.

Mossy. Term applied to emeralds clouded by fissures.

Muddy. Imperfect crystallisation which obstructs the passage of light; exemplified by mud stirred in water.

Muffle. An oven-shaped vessel of baked fire-clay containing cupels or cups in which alloy is fused, or a furnace with a chamber surrounded by incandescent fuel.

Mytilidae. A family of conchiferous molluscs—pearl producing mussels.

Mytilius Edulis. The true mussel.

Naats. Thin flat crystals (diamond) used for "roses" and, by resplitting, for draw-plates.

Nacreous. Lustre resembling mother-of-pearl, the lining of mollusc shells.

Night Emerald. Olivine, which loses its yellow tint by artificial light, showing only its green.

Noble. The highest type of a specified kind of gem, as "Noble Opal." A synonym of "Precious."

Nodules. A rounded irregular-shaped lump or mass, sometimes enclosing a foreign body in the centre.

Occurrence. To be found existing.

Octahedron. Two four-sided pyramids united base to base.

Off Colour. Having but a tint of desirable colour.

Old Mine. Diamonds from the old Brazilian fields; old cut diamonds of good colour.

Opacity. The quality or state of being impervious to light.

Opalescence. A milky or pearly reflection from the interior of a stone.

Opalescent. Resembling or having the tints of opal; reflecting lustre from a single spot.

Opaque. When no light is transmitted.

Optic Axis. The line in a double refracting crystal in the direction of which no double refraction occurs.

Organic. Pertaining to the animal or vegetable kingdom.

Oriental. A term much used in the gem trade to distinguish stones of entirely differing chemical and crystallographic nature to which a common name is applied, as "Oriental topaz," bestowed on specimens of yellow corundum of gem quality.

Original Lots. Unbroken parcels of diamonds as graded and assorted at the mines.

Orthorhombic. (Trimetric.) Having three unequal axes intersecting at right angles.

Oxide. The product of the combination of oxygen with a metal or metalloid.

Panning. Primitive process of washing gravel by placer miners in search for gems.

Pearly. Resembling the sheen of the pearl.

Percussion. (Shaking Table.) A form of ore-separating apparatus consisting of a slightly sloping table on which stamped ore or metalliferous sand is placed to be sorted by gravity. A stream of water is directed over the ore, and the table is subjected to concussion at intervals.

Phosphorescence. The property possessed by substances of emitting light in certain conditions.

Pigeon Blood. A deep clear red; the gem colour of the most highly prized specimens of the ruby.

Placer. A deposit of gem minerals found separately, sometimes as rolled pebbles, in alluvium or diluvium, or beds of streams.

Play of Colours. (See colour-play.)

Pleochrism. The term applied to minerals in which a different shade of colour is seen in more than two directions.

Polarisation. In optics, a state into which the ethereal undulations which cause the sensation of light are brought under certain conditions.

Pomegranate. Translation of the Hindu name for spinel.

Precious. (See "Noble.")

Primary Situation. A mineral found in the rock in which it was formed.

Prism. (Geometry.) A solid having similar and parallel bases, its sides forming similar parallelograms. (Optics.) Any transparent medium comprised between plane faces, usually inclined to each other.

Prospecting. Searching for gem fields or mines.

Pulsator. (See Jig; Jigger.)

Pyroelectric. (Thermo-electric.) Pertaining or relating to electric currents or effects produced by heat.

Quality. Native values of a gem irrespective of colour and cut.

Reconstructed. A term applied to an artificial gem composed of fused particles of a natural precious stone—"Reconstructed rubies" although not difficult to differentiate by tests, from the red corundum of gem quality from Nature's laboratory, attain some commercial success. Also called "Scientific Ruby."

Reflection. The act of reflecting or throwing back, as of rays of light.

Refraction. Bending back. In optics, the refraction of a ray of light into a number of other rays forming a hollow cone. Double Refraction: In crystals that are not homogeneous but have different properties of elasticity, etc., in different directions, if a ray of light enter the crystal in some particular directions it is not simply refracted but divided into two rays.

REJECTIONS. Diamonds not worthy of cutting.

RENIFORM. Kidney-shaped.

RESINOUS. The lustre of yellow resins; manifested in the common forms of garnets.

RHOMBS. Lozenge-shaped faces.

RIVERS. Diamonds found in the beds of rivers.

RÖNTGEN RAYS. (See X-rays.)

ROSETTE. (Rose-cut.) A form of cutting in which the stone's base is a single face; the general form is pyramidal and the several varieties each possess a different number of facets; a Double Rosette, also called "Pendeloque" is of the form of two rosettes joined at their bases.

ROUGH. Uncut crystals.

ROUND-STONES. Diamond crystals with arched facets.

SCHIST. A term used for rocks consisting of mineral ingredients arranged so as to impart a more or less laminar structure that may be broken into slabs or slaty fragments.

SECOND BYE. Fifth grade of rough diamonds.

SECOND CAPE. Third grade of South African rough diamonds.

SEMITRANSPARENT. When objects are visible through a mineral, though the outlines are indistinct.

SHARPS. Thin, knife-edge pieces of diamond.

SIAMS. Dark, garnet-coloured rubies usually found in Siam.

SIGHT. Exhibition of rough diamonds by the London Syndicate to applicants for the privilege of inspecting and purchasing.

SILK. White, glistening streaks in the grain of rubies.

SILKY. A lustre suggesting silk, as exhibited by crocidolite.

SILVER CAPES. Diamonds having a very slight tint of yellow.

Skip. A bucket employed in narrow or inclined mine shafts, where the hoisting device must be confined between guides.

Smaragdus. Ancient name for emerald and other green stones.

Sorters. The experts at the South African diamond mines who assort the rough diamonds.

Sorting Tables. Tables on which rough diamonds are assorted.

Specific Gravity. The relative weight of bulk as compared with distilled water at 60° F.

Spectrum. The coloured image or images produced when the rays from any source of light are decomposed or dispersed by refraction through a prism.

Splints. Thin, pointed pieces of diamonds.

Spread. Surface in proportion to the depth of a stone.

Star Stones. Sapphires, and sometimes rubies, which by structure and cutting are seen to be asteriated, exhibiting a star of six rays of light.

Step-Cut. (Trap-Cut.) A form of cutting employed for stones not deeply coloured when they are not cut as brilliants; a simple typical form is that of a stepped pyramid with the apex sliced off.

Streak. Colour of the surface of a stone after being rubbed or scratched. "Streak-Powder" is the powder abraded from a stone.

Striated. Term applied to minerals which exhibit lines traversing the plane of a crystal; such lines bear a definite relation to certain forms of the mineral on which they occur.

Subtranslucent. When the edges of a mineral only transmit light faintly.

Table-Stone. The typical form thus described is a style of diamond-cutting derived from an octahedron by cutting to opposite corners to an equal amount.

Tailings. The refuse part of washed gem ground, rock, or gravel which is thrown behind the tail of the washing apparatus and which is put through a second process to recover values possibly remaining.

TALCOSE. Partaking of the characters of talc.

TALLOW-TOPPED. A stone cut with a flattish convex surface.

TARIFF. Ten per cent. import duty imposed upon cut diamonds by the United States Government.

TETRAGONAL. Pertaining to a tetragon; having four angles or sides, as a square, quadrangle, or rhomb.

TETRAGONAL SYSTEM. A system of crystallisation in which the lateral axes are equal, being the diameters of a square, while the vertical is either longer or shorter than the lateral. Called also Dimetric, Monadimetric, or Pyramidal System.

TIFFANYITE. A hydrocarbon, causing phosphorescence and opalescence in some precious stones.

TOP CRYSTALS. Standard grade of diamonds.

TORN END. A three-cornered pyramid from the point of a wassie.

TRANSLUCENT. Minerals so nearly opaque that objects are scarcely, if at all, visible through them.

TRANSPARENT. When the outlines of an object can be seen through a gem distinctly.

TRICLINIC. The system in crystallography in which the three crystallographic axes are unequal, and inclined at angles which are not right angles, so that the forms are oblique in every direction, and have no plane of symmetry.

TWINNED. Two or more distinct crystals which have been formed in conjunction.

UNIAXIAL. Having one direction within the crystal, along which a ray of light can proceed without being bifurcated.

UNIO. The river mussel; the type-genus of Unionidæ, with more than 400 species from all parts of the world.

URALIAN. Minerals from the Ural Mountains, Siberia.

VITREOUS. Glassy, as glassy lustre.

WASSIE. A large cleavage of a crystal split for cutting, as an octahedron divided into two pieces.

WAXY. A distinctive lustre, as of the turquoise.

WEATHERING. The disintegration and decay of minerals under the influence of the weather.

Well. Name given to the dark centre of a diamond cut too thick.

Wesseltons. Third grade cut diamonds.

X-Rays. (Röntgen Rays.) A recently discovered form of radiant energy that is sent out when the cathode rays of a Crookes tube strike upon the opposite walls of the tube or upon any object in the tube; discovered in 1895 at Würzburg, Germany, by Professor W. C. Röntgen. By means of these rays it is possible to see and photograph bones, bullets, or other opaque objects through the fleshy parts of the body. The X-rays are of some value in testing mineral substances represented as precious stones. Under X-rays the diamond is transparent; the glass, or "strass," used to manufacture imitation diamonds is always opaque under this exposure.

Yellow Ground. The upper diamond-bearing clay of South African mines.

INDEX

Q

R

S

Diamonds

A Study of the Factors that Govern their Value

By

Frank B. Wade

"I shall speak a little more of the diamonds, that they who know them not may not be deceived by chapmen who go through the country selling them, for whoever will buy the diamond, it is needful that he know them, . . ." —Chap. XIV., *The Voyages and Travels of Sir John Maundeville.*

Table of Contents

G. P. Putnam's Sons

New York London

The Magic and Science of Jewels and Stones

By Isidore Kozminsky

This book presents, in attractive literary form, the ideas of the ancients and moderns in regard to the use of precious gems. It explains, for instance, the import of the breastplate, and analyzes the history and the meaning of various legends, stories, and parables connected with gems.

It gives an interesting account of the use of gems as symbols, birthstones, talismans, and stones of fortune. It gives a dramatic description of the gems that have been famous in history: diamonds, rubies, sapphires, pearls, and opals, and particularly of the wonderful "Flame Queen."

A chapter is devoted to the science, literature, and the poetry associated with gems.

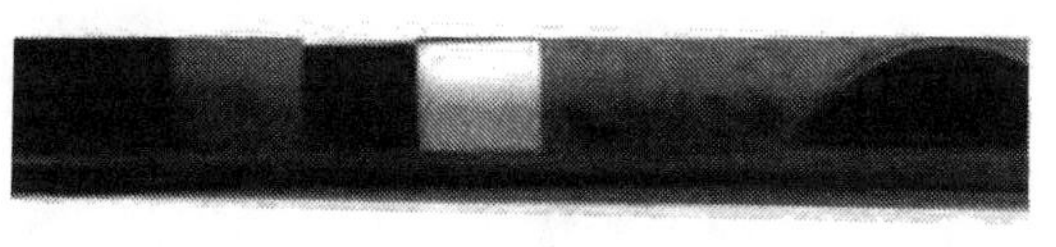

CPSIA information can be obtained
at www.ICGtesting.com
Printed in the USA
BVHW041346241120
594109BV00017B/153

9 780343 859923